Modern Raman Spectroscopy
– A Practical Approach

Modern Raman Spectroscopy – A Practical Approach

Ewen Smith
Strathclyde University, Glasgow

Geoffrey Dent
Intertek ASG and UMIST, Manchester

John Wiley & Sons, Ltd

Other Wiley Editorial Offices

John Wiley & Sons, Inc., 111 River Street, Hoboken, NJ 07030, USA

Jossey-Bass, 989 Market Street, San Francisco, CA 94103-1741, USA

Wiley-VCH Verlag GmbH, Boschstr. 12, D-69469 Weinheim, Germany

John Wiley & Sons Australia Ltd, 33 Park Road, Milton, Queensland 4064, Australia

John Wiley & Sons (Asia) Pte Ltd, 2 Clementi Loop #02-01, Jin Xing Distripark, Singapore 129809

John Wiley & Sons Canada Ltd, 22 Worcester Road, Etobicoke, Ontario, Canada M9W 1L1

Wiley also publishes its books in a variety of electronic formats. Some content that appears in print may
not be available in electronic books.

Library of Congress Cataloging in Publication Data

Smith, Ewen.
    Modern Raman spectroscopy : a practical approach / Ewen Smith, Geoff Dent.
        p.   cm.
    Includes bibliographical references and index.
    ISBN 0-471-49668-5 (cloth : alk. paper) — ISBN 0-471-49794-0 (pbk. : alk. paper)
    1. Raman spectroscopy.   I. Dent, Geoffrey.   II. Title.
    QD96.R34S58 2005
    535.8′46—dc22
                                                                    2004014375

British Library Cataloguing in Publication Data

A catalogue record for this book is available from the British Library

ISBN-13 978-0-471-49794-3 (P\B)
ISBN-13 978-0-471-49668-7 (H\B)

Typeset in 10/12 pt Times by Integra Software Services Pvt. Ltd, Pondicherry, India

Contents

Preface

For many years the practice of Raman spectroscopy was confined to experts in dedicated academic or industrial research laboratories. The instruments were large, complicated and the experiments could be quite complex. With advances in modern technology, Raman spectrometers have become small, portable and are regularly used by people who are neither specialist spectroscopists nor analysts. Often instruments are bought for a specific application but eventually the user asks, 'What else can this be used for?'. Whilst much good work continues to be carried on by Raman experts in rolling back the frontiers in advanced techniques, this book is addressed to the more general, modern, application-driven user. Our aim in writing this book is to provide the information necessary to enable new users to understand and apply the technique correctly. This includes descriptions of the many pitfalls that can be encountered. We wish to aid those with a more sustained interest to gain sufficient knowledge and understanding to make full use of the high information content that Raman scattering can afford. With this approach in mind, we have provided in the early chapters enough basic theory to make a practical interpretation of Raman spectra. The theory is dealt with in a little more depth in later chapters where the approach is to describe the main equations used to explain Raman scattering, but concentrating on their meaning and relevance rather than a full mathematical treatment.

With this background the much more detailed world is revealed in which aspects of Raman spectroscopy can provide unique information for a limited number of analytical problems. A full mathematical approach to the theory of Raman spectroscopy is outside the scope of this book. For those who read through to the end, the book will provide a firm grounding, with appropriate references given, from which to approach more in-depth studies of specific aspects of Raman spectroscopy. In writing this book some difficult choices have had to be made particularly around the presentation of the theory. Many current users of Raman spectroscopy have little idea of the underlying modern theory and as a result are at risk of misinterpreting their results. However, whilst a full explanation of theory has to have some mathematics, in the authors' experience many users do not have the time or the background to understand a fully rigorous mathematical exposition. The non-rigorous mathematical approach is almost essential. We have used as few equations as possible to show how the theory is developed and those are deliberately not in the first chapter. The equations are explained rather than derived so that those with little knowledge of mathematics can understand the physical meaning described. This level of understanding is sufficient for most purposes. Where a more in-depth approach is sought, the explanation would serve as a good starting point. Two theories are often used in Raman spectroscopy – classical theory and quantum theory. A consequence of our approach to the theory is the omission of classical Raman theory altogether. Classical theory does not use quantum mechanics. In the authors' opinion the lack of quantum theory to describe vibrations means that it does not deliver the information required by the average Raman spectroscopist.

One of the practical difficulties faced is in compliance with the IUPAC convention in the description of spectrum scales. Whilst the direction of the wavenumber shift should always be consistent, this is not the practice in most scientific journals or by software writers for instrument companies. Unfortunately the modern practitioner has to view original and reference spectra in differing formats. To illustrate applications we have used the format in which the user is most likely to see a reference spectrum. Equally, where we have used, with permission, literature examples in this book, it would not be possible to change these round to fit the convention. Raman scattering is a shift from an exciting frequency and should be labelled $\Delta\,cm^{-1}$. However it is common practice to use cm^{-1} with the delta implied. Changing labels on previously published examples would not be permitted so for simplicity and consistency we have used the common format. We apologise to the purists who would prefer complete compliance with the IUPAC convention, but we have found that it is not practicable.

It is the authors' hope that those who are just developing or reviving an interest in Raman spectroscopy will very quickly gain a practical understanding from the first two chapters. Furthermore they will be inspired by the elegance and information content of the technique to delve further into the rest of the book, and explore the vast potential of the more sophisticated applications of Raman spectroscopy.

Acknowledgements

We thank members of Professor Smith's group at the University of Strathclyde who read chapters and supplied diagrams – Rachael Littleford, Graeme McNay, Prokopis Andrikopoulos, Dale Cunningham and Maarten Scholtes; Gillian Neeson and Liz Keys who did much of the typing; Intertek ASG for permisson to publish; and our respective wives, Frances and Thelma, for putting up with us.

Chapter 1

Introduction, Basic Theory and Principles

1.1 INTRODUCTION

The main spectroscopies employed to detect vibrations in molecules are based on the processes of infrared absorption and Raman scattering. They are widely used to provide information on chemical structures and physical forms, to identify substances from the characteristic spectral patterns ('fingerprinting'), and to determine quantitatively or semi-quantitatively the amount of a substance in a sample. Samples can be examined in a whole range of physical states; for example, as solids, liquids or vapours, in hot or cold states, in bulk, as microscopic particles, or as surface layers. The techniques are very wide ranging and provide solutions to a host of interesting and challenging analytical problems. Raman scattering is less widely used than infrared absorption, largely due to problems with sample degradation and fluorescence. However, recent advances in instrument technology have simplified the equipment and reduced the problems substantially. These advances, together with the ability of Raman spectroscopy to examine aqueous solutions, samples inside glass containers and samples without any preparation, have led to a rapid growth in the application of the technique.

In practice, modern Raman spectroscopy is simple. Variable instrument parameters are few, spectral manipulation is minimal and a simple interpretation of the data may be sufficient. This chapter and Chapter 2 aim to set out the basic principles and experimental methods to give the reader a firm understanding of the basic theory and practical considerations so that the technique

Modern Raman Spectroscopy – A Practical Approach   W.E. Smith and G. Dent
© 2005 John Wiley & Sons, Ltd   ISBNs: 0-471-49668-5 (HB); 0-471-49794-0 (PB)

can be applied at the level often required for current applications. However, Raman scattering is an underdeveloped technique, with much important information often not used or recognized. Later chapters will develop the minimum theory required to give a more in-depth understanding of the data obtained and to enable comprehension of some of the many more advanced techniques which have specific advantages for some applications.

1.1.1 History

The phenomenon of inelastic scattering of light was first postulated by Smekal in 1923 [1] and first observed experimentally in 1928 by Raman and Krishnan [2]. Since then, the phenomenon has been referred to as Raman spectroscopy. In the original experiment sunlight was focussed by a telescope onto a sample which was either a purified liquid or a dust-free vapour. A second lens was placed by the sample to collect the scattered radiation. A system of optical filters was used to show the existence of scattered radiation with an altered frequency from the incident light – the basic characteristic of Raman spectroscopy.

1.2 BASIC THEORY

When light interacts with matter, the photons which make up the light may be absorbed or scattered, or may not interact with the material and may pass straight through it. If the energy of an incident photon corresponds to the energy gap between the ground state of a molecule and an excited state, the photon may be absorbed and the molecule promoted to the higher energy excited state. It is this change which is measured in absorption spectroscopy by the detection of the loss of that energy of radiation from the light. However, it is also possible for the photon to interact with the molecule and scatter from it. In this case there is no need for the photon to have an energy which matches the difference between two energy levels of the molecule. The scattered photons can be observed by collecting light at an angle to the incident light beam, and provided there is no absorption from any electronic transitions which have similar energies to that of the incident light, the efficiency increases as the fourth power of the frequency of the incident light.

Scattering is a commonly used technique. For example, it is widely used for measuring particle size and size distribution down to sizes less than 1 μm. One everyday illustration of this is that the sky is blue because the higher energy blue light is scattered from molecules and particles in the atmosphere more efficiently than the lower energy red light. However, the main scattering technique used for molecular identification is Raman scattering.

The process of absorption is used in a wide range of spectroscopic techniques. For example it is used in acoustic spectroscopy where there is a very small energy difference between the ground and excited states and in X-ray absorption spectroscopy where there is a very large difference. In between these extremes are many of the common techniques such as NMR, EPR, infrared absorption, electronic absorption and fluorescence emission, and vacuum ultraviolet (UV) spectroscopy. Figure 1.1 indicates the wavelength ranges of some commonly used types of radiation.

Radiation is often characterized by its wavelength (λ). However, in spectroscopy, because we are interested in the interaction of radiation with states of the molecule being examined and this being usually discussed in terms of energy, it is often useful to use frequency (ν) or wavenumber (ϖ) scales, which are linearly related with energy. The relationships between these scales are given below:

$$\lambda = c/\nu \qquad (1.1)$$

$$\nu = \Delta E/h \qquad (1.2)$$

$$\varpi = \nu/c = 1/\lambda \qquad (1.3)$$

It is clear from Equations (1.1)–(1.3) that the energy is proportional to the reciprocal of wavelength and therefore the highest energy region is on the left in Figure 1.1 and the longest wavelength on the right.

The way in which radiation is employed in infrared and Raman spectroscopies is different. In infrared spectroscopy, infrared energy covering a range of frequencies is directed onto the sample. Absorption occurs where the frequency of the incident radiation matches that of a vibration so that the molecule is promoted to a vibrational excited state. The loss of this frequency of radiation from the beam after it passes through the sample is then detected. In contrast, Raman spectroscopy uses a single frequency of radiation to irradiate the sample and it is the radiation scattered from the molecule, one vibrational unit of energy different from the incident beam, which is detected. Thus, unlike infrared absorption, Raman scattering does not require matching of the incident radiation to the energy difference between the ground and excited states. In Raman scattering, the light interacts with the molecule and distorts (polarizes) the cloud of electrons round the nuclei to form a short-lived state

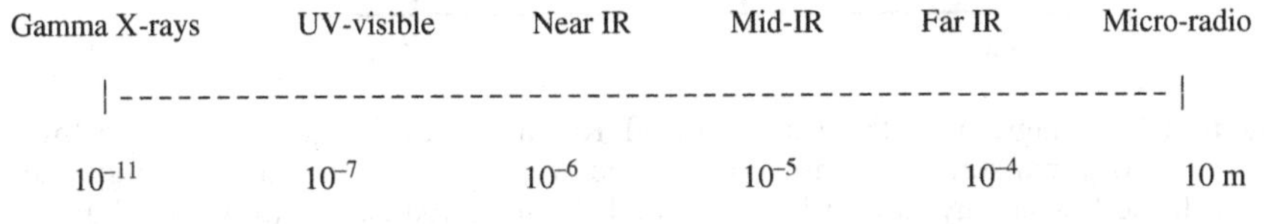

Gamma X-rays	UV-visible	Near IR	Mid-IR	Far IR	Micro-radio
10^{-11}	10^{-7}	10^{-6}	10^{-5}	10^{-4}	10 m

Figure 1.1. The electromagnetic spectrum on the wavelength scale in metres.

called a 'virtual state', which is discussed in Chapter 3. This state is not stable and the photon is quickly re-radiated.

The energy changes we detect in vibrational spectroscopy are those required to cause nuclear motion. If only electron cloud distortion is involved in scattering, the photons will be scattered with very small frequency changes, as the electrons are comparatively light. This scattering process is regarded as elastic scattering and is the dominant process. For molecules it is called Rayleigh scattering. However, if nuclear motion is induced during the scattering process, energy will be transferred either from the incident photon to the molecule or from the molecule to the scattered photon. In these cases the process is inelastic and the energy of the scattered photon is different from that of the incident photon by one vibrational unit. This is Raman scattering. It is inherently a weak process in that only one in every 10^6–10^8 photons which scatter is Raman scattered. In itself this does not make the process insensitive since with modern lasers and microscopes, very high power densities can be delivered to very small samples but it is does follow that other processes such as sample degradation and fluorescence can readily occur.

Figure 1.2 shows the basic processes which occur for one vibration. At room temperature, most molecules, but not all, are present in the lowest energy vibrational level. Since the virtual states are not real states of the molecule but are created when the laser interacts with the electrons and causes polarization, the energy of these states is determined by the frequency of the light source used. The Rayleigh process will be the most intense process since most photons scatter this way. It does not involve any energy change and consequently the light returns to the same energy state. The Raman scattering process from the ground vibrational state m leads to absorption of energy by the molecule and its promotion to a higher energy excited vibrational state (n). This is called Stokes

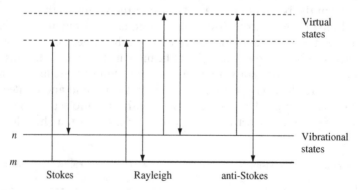

Figure 1.2. Diagram of the Rayleigh and Raman scattering processes. The lowest energy vibrational state m is shown at the foot with states of increasing energy above it. Both the low energy (upward arrows) and the scattered energy (downward arrows) have much larger energies than the energy of a vibration.

scattering. However, due to thermal energy, some molecules may be present in an excited state such as *n* in Figure 1.2. Scattering from these states to the ground state *m* is called anti-Stokes scattering and involves transfer of energy to the scattered photon. The relative intensities of the two processes depend on the population of the various states of the molecule. The populations can be worked out from the Boltzmann equation (Chapter 3) but at room temperature, the number of molecules expected to be in an excited vibrational state other than any really low-energy ones will be small.

Thus, compared to Stokes scattering, anti-Stokes scattering will be weak and will become weaker as the frequency of the vibration increases, due to decreased population of the excited vibrational states. Further, anti-Stokes scattering will increase relative to Stokes scattering as the temperature rises. Figure 1.3 shows a typical spectrum of Stokes and anti-Stokes scattering from cyclohexane separated by the intense Rayleigh scattering which should be off-scale close to the point where there is no energy shift. However there is practically no signal close to the frequency of the exciting line along the *x*-axis. This is because filters in front of the spectrometer remove almost all light within about 200 cm^{-1} of the exciting line. Some breakthrough of the laser light can be seen where there is no energy shift at all.

Usually, Raman scattering is recorded only on the low-energy side to give Stokes scattering but occasionally anti-Stokes scattering is preferred. For example, where there is fluorescence interference, this will occur at a lower energy than the excitation frequency and consequently anti-Stokes scattering can be used to avoid interference. The difference in intensities of Raman bands in Stokes and anti-Stokes scattering can also be used to measure temperature.

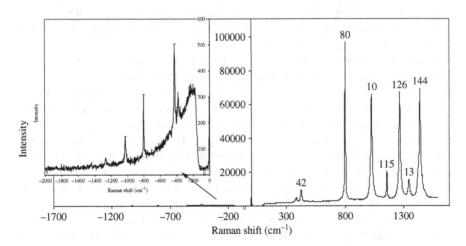

Figure 1.3. Stokes and anti-Stokes scattering for cyclohexane. To show the weak anti-Stokes spectrum, the *y*-axis has been extended in the inset.

Figure 1.2 illustrates one key difference between infrared absorption and Raman scattering. As described above, infrared absorption would involve direct excitation of the molecule from state m to state n by a photon of exactly the energy difference between them. In contrast, Raman scattering uses much higher energy radiation and measures the difference in energy between n and m by subtracting the energy of the scattered photon from that of the incident beam (the two vertical arrows in each case).

The cyclohexane spectrum in Figure 1.3 shows that there is more than one vibration which gives effective Raman scattering (i.e. is Raman active); the nature of these vibrations will be discussed in Section 1.3. However, there is a basic selection rule which is required to understand this pattern. Intense Raman scattering occurs from vibrations which cause a change in the polarizability of the electron cloud round the molecule. Usually, symmetric vibrations cause the largest changes and give the greatest scattering. This contrasts with infrared absorption where the most intense absorption is caused by a change in dipole and hence asymmetric vibrations which cause this are the most intense. As will be seen later, not all vibrations of a molecule need, or in some cases can, be both infrared and Raman active and the two techniques usually give quite different intensity patterns. As a result the two are often complementary and, used together, give a better view of the vibrational structure of a molecule.

One specific class of molecule provides an additional selection rule. In a centrosymmetric molecule, no band can be active in both Raman scattering and infrared absorption. This is sometimes called the mutual exclusion rule. In a centrosymmetric molecule, reflection of any point through the centre will reach an identical point on the other side (C_2H_4 is centrosymmetric, CH_4 is not). This distinction is useful particularly for small molecules where a comparison of the spectra obtained from infrared absorption and Raman scattering can be used to differentiate *cis* and *trans* forms of a molecule in molecules such as a simple azo dye or a transition metal complex.

Figure 1.4 shows a comparison of the infrared and Raman spectra for benzoic acid. The x-axis is given in wavenumbers for which the unit is cm^{-1}. Wavenumbers are not recommended SI units but the practice of spectroscopy is universally carried out using these and this is unlikely to change. For infrared absorption each peak represents an energy of radiation absorbed by the molecule. The y-axis gives the amount of the light absorbed and is usually shown with the maximum absorbance as the lowest point on the trace. Raman scattering is presented only as the Stokes spectrum and is given as a shift in energy from the energy of the laser beam. This is obtained by subtracting the scattered energy from the laser energy. In this way the difference in energy corresponding to the ground and excited vibrational states (n and m in Figure 1.2) is obtained. This energy difference is what is measured directly by infrared. The scattering is measured as light detected by the spectrometer and the maximum amount of light detected is the highest point on the trace.

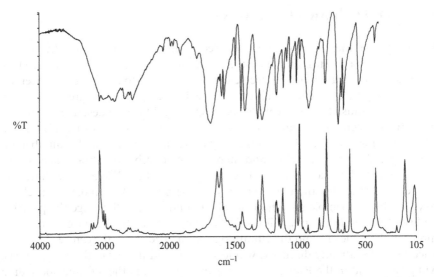

%T

4000 3000 2000 1500 1000 500 105

cm⁻¹

Figure 1.4. Infrared and Raman spectra of benzoic acid. The top trace is infrared absorption given in % transmission (%T) so that the lower the transmission value the greater the absorption. The lower trace is Raman scattering and the higher the peak the greater the scattering.

Strictly speaking, Raman scattering should be expressed as a shift in energy from that of the exciting radiation and should be referred to as $\Delta\,cm^{-1}$ but it is often expressed simply as cm^{-1}. This practice is followed in this book for simplicity. Although different energy ranges are possible, the information of interest to most users is in the $3600{-}400\,cm^{-1}$ (2.8–12 micron) range in infrared spectroscopy and down to $200\,cm^{-1}$ in Raman spectroscopy since this includes most modes which are characteristic of a molecule. In some applications, much larger or smaller energy changes are studied and modern Raman equipment can cope with much wider ranges. One specific advantage of Raman scattering is that shifts from the laser line of $50\,cm^{-1}$ or lower can easily be recorded with the correct equipment. Many modern machines for reasons of cost and simplicity are not configured in a suitable way to measure shifts below $100{-}200\,cm^{-1}$. The intensities of the bands in the Raman spectrum are dependent on the nature of the vibration being studied and on instrumentation and sampling factors. Modern instruments should be calibrated to remove the instrument factors but this is not always the case; these factors are dealt with in the next chapter. Sampling has a large effect on the absolute intensities, bandwidths observed and band positions. Again these will be dealt with later. This chapter will concentrate on the effect on Raman scattering of the set of vibrations present in a molecule and set out a step-by-step approach to interpretation based on simple selection rules.

1.3 MOLECULAR VIBRATIONS

Provided that there is no change in electronic energy, for example, by the absorption of a photon and the promotion of an electron to an excited electronic state, the energy of a molecule can be divided into a number of different parts or 'degrees of freedom'. Three of these degrees of freedom are taken up to describe the translation of the molecule in space and three to describe rotational movement except for linear molecules where only two types of rotation are possible. Thus, if N is the number of atoms in a molecule, the number of vibrational degrees of freedom and therefore the number of vibrations possible is $3N - 6$ for all molecules except linear ones where it is $3N - 5$. For a diatomic molecule, this means there will be only one vibration. In a molecule such as oxygen, this is a simple stretch of the O–O bond. This will change the polarizability of the molecule but will not induce any dipole change since there is no dipole in the molecule and the vibration is symmetric about the centre. Thus the selection rules already discussed would predict, and it is true, that oxygen gas will give a band in the Raman spectrum and no band in the infrared spectrum. However in a molecule such as nitric oxide, NO, there will be only one band but, since there is both a dipole change and a polarizability change, it will appear in both the infrared and Raman spectrum.

A triatomic molecule will have three modes of vibration. They are a symmetrical stretch, a bending or deformation mode and an asymmetrical stretch as shown in Figure 1.5. The very different water (H_2O) and carbon dioxide (CO_2) molecules clearly demonstrate these vibrations. These diagrams use 'spring and ball' models. The spring represents the bond or bonds between the atoms. The stronger the bond the higher the frequency. The balls represent the atoms and the heavier they are the lower the frequency. The expression which relates the mass of the atoms and the bond strength to the vibrational frequency is Hooke's

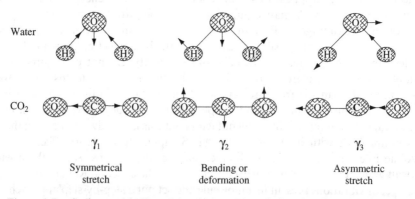

Figure 1.5. Spring and ball model – three modes of vibration for H_2O and CO_2.

law which is dealt with in Chapter 3, but for the present, it is clear that strong bonds and light atoms will give higher frequencies of vibration and heavy atoms and weak bonds will give lower ones.

This simple model is widely used to interpret vibrational spectra. However, the molecule actually exists as a three-dimensional structure with a pattern of varying electron density covering the whole molecule. A simple depiction of this for carbon dioxide is shown in Figure 1.6. If either molecule vibrates, the electron cloud will alter as the positive nuclei change position and depending on the nature of the change, this can cause a change of dipole moment or polarization. In these triatomic molecules, the symmetrical stretch causes large polarization changes and hence strong Raman scattering with weak or no dipole change and hence weak or no infrared absorption. The deformation mode causes a dipole change but little polarization change and hence strong infrared absorption and weak or non-existent Raman scattering.

As an example of this, Figure 1.7 illustrates the vibrations possible for carbon disulphide along with the corresponding infrared absorption and Raman scattering spectra.

Although this type of analysis is suitable for small molecules, it is more difficult to apply in a more complex molecule. Figure 1.8 shows one vibration from a dye in which a large number of atoms are involved. This is obtained from a theoretical calculation using density functional theory (DFT) which is discussed briefly later. It probably gives a depiction of the vibration which is close to the truth. However, even if it were possible to calculate the spectrum of every molecule quickly in the laboratory, which at present it is not, this type of diagram is only of limited utility to the spectroscopist. A comparison between molecules of similar type is difficult unless a full calculation is available for them all and each subtle change in the nuclear displacements is drawn out or accurately described for each one. This limits the ability to compare large numbers of molecules or to understand the nature of vibrations in molecules for which there is no calculation.

The usual approach to describing vibrations is to simplify the problem and break the displacements down into a number of characteristic features, which

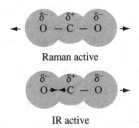

Figure 1.6. Electron cloud model of water and carbon dioxide showing an IR and a Raman active vibrations.

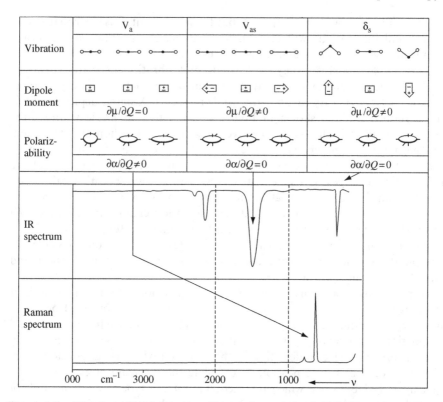

Figure 1.7. Dipole and polarization changes in carbon disulphide, with resultant infrared and Raman spectra. (Reprinted from A. Fadini and F.-M. Schnepel, *Vibrational Spectroscopy: Methods and Applications*, Ellis Horwood Ltd, Chichester, 1989.)

can relate to more than one molecule. In the vibration in Figure 1.8 which comes from a calculation to predict the energies of vibrations each azo dye, the biggest displacements of the heavier atoms is on one of the ring systems. The vibration would almost certainly be labelled vaguely as a 'ring stretch'. In another vibration not shown the situation was much simpler. Large displacements were found on the two nitrogen atoms which form the azo bond between the rings, and the direction indicated bond lengthening and contracting during the vibrational cycle. Thus this vibration is called the azo stretch, and there is a change in polarizability just as there was for oxygen; so it should be a Raman-active vibration. We can search for these vibrations in the actual spectrum and hopefully match a peak to the vibration. This is called assigning the vibration. Thus, it is possible to describe a vibration in a few helpful words. In some cases this is fairly accurate as for the azo stretch, but in some cases, the description is not adequate to describe the actual movement. However, common bands can be assigned and reasonably described in many molecules.

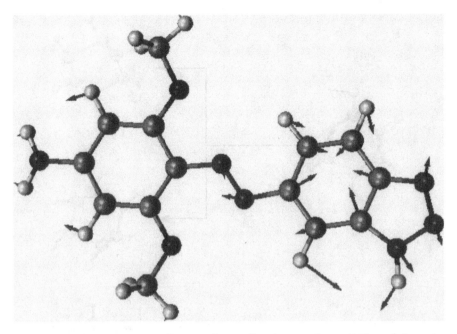

Figure 1.8. A displacement diagram for a vibration at about $1200\,cm^{-1}$ in a dye indicating the involvement of a number of atoms. The arrows show the direction of the displacement. Since the equilibrium position of the atoms is shown, during a complete vibration the arrows will reverse in direction.

1.3.1 Group Vibrations

To assign vibrations to spectral peaks it is necessary to realize that two or more bonds which are close together in a molecule and are of similar energies can interact and it is the vibration of the group of atoms linked by these bonds which is observed in the spectrum. For example, the CH_2 group is said to have a symmetric and an anti-symmetric stretch rather than two separate CH stretches (Figure 1.9). It follows from this and from the geometry of the molecule that different types of vibrations are possible for different groups. Selected examples of a few of these for CH_3 and C_6H_6 are shown in Figure 1.9.

In contrast, where there is a large difference in energy between the vibrations in different bonds or if the atoms are well separated in the molecule, they can be treated separately. Thus, for CH_3Br, the C–H bonds in CH_3 must be treated as a group but the C–Br vibration is treated separately. In Figure 1.9, the selected vibrations of benzene are shown in two different ways. Firstly they are shown with the molecule in the equilibrium position with arrows showing the direction of the vibrational displacement. To illustrate what this means, they are also shown with the vibration at the extremes of the vibrational movement. To show

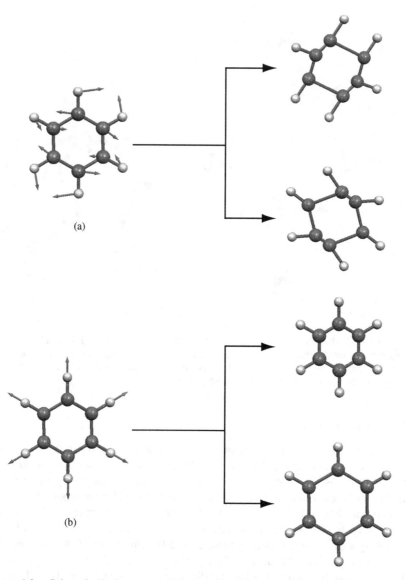

Figure 1.9. Selected displacement diagrams for benzene and for CH_3 in CH_3Br. (a) A quadrant stretch for benzene at about $1600\,\mathrm{cm}^{-1}$. (b) The symmetric breathing mode at just above $1000\,\mathrm{cm}^{-1}$. (c and d) Two C–H vibrations at about $3000\,\mathrm{cm}^{-1}$. (e) The symmetric stretch of CH_3 in CH_3Br at above $3000\,\mathrm{cm}^{-1}$. (f) An asymmetric stretch at above $3000\,\mathrm{cm}^{-1}$. (g) A CH bend at about $1450-1500\,\mathrm{cm}^{-1}$. (h) A low frequency mode at below $600\,\mathrm{cm}^{-1}$.

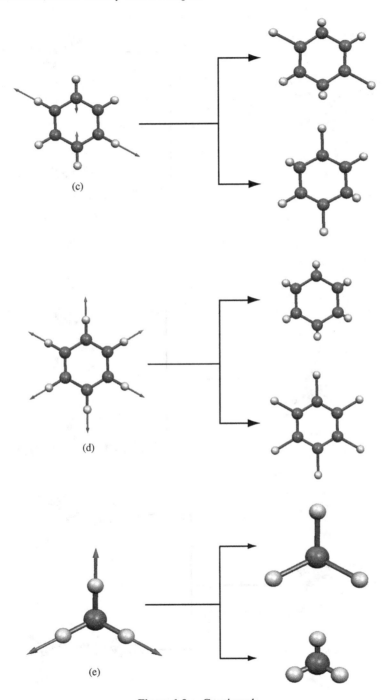

Figure 1.9. Continued.

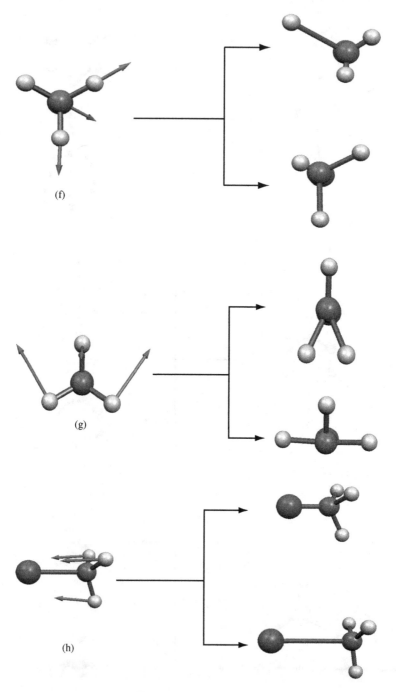

Figure 1.9. Continued.

the selected CH_3 group vibrations, the molecule is completed using a bromine. As discussed, the C–Br bond vibrates at a much lower frequency and does not interact appreciably with the high CH_3 displacements shown.

1.3.2 An Approach to Interpretation

It is possible to give energy ranges in which the characteristic frequencies of the most common groups which are strong in either infrared or Raman scattering can occur. The relative intensities of specific peaks help to confirm that the correct vibration has been picked out.

For example, carbonyl groups $>C=O$ which are both asymmetric and ionic will have a dipole moment which will change when the group stretches in a manner analogous to oxygen. They have strong bands in the infrared spectrum

Table 1.1. Single vibration and group frequencies and possible intensities of peaks commonly identified in Raman scattering. The length of the vertical line represents the wavenumber range in cm^{-1} in which each type of vibration is normally found and the line thickness gives an indication of intensity with thick lines being the most intense.

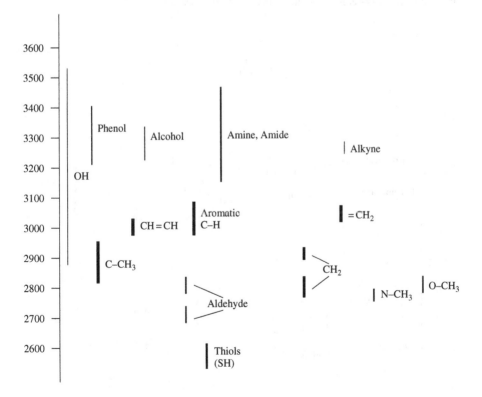

but are weaker in the Raman spectrum. They are usually present at $\sim$1700 cm^{-1}. Symmetrical groups such as unsaturated bonds (–C=C–) and disulphide bonds (–S–S–) are weak infrared absorbers, but strong Raman scatterers. The stretching modes for these vibrations are $\sim$1640 and 500 cm^{-1} respectively. There are many more examples. It is the combination of the knowledge of approximate energy and likely relative intensity of particular vibrations which form the basis of the assignment mode used by most spectroscopists. For example, the 4000–2500 cm^{-1} is the region where single bonds (X–H) absorb. The 2500–2000 cm^{-1} is referred to as the multiple bond (–N=C=O) region. The 2000–1500 cm^{-1} region is where double bonds (–C=O, –C=N, –C=C–) occur. Below 1500 cm^{-1}, some groups, e.g. nitro (O=N=O) do have specific bands but many molecules have complex patterns of Carbon–Carbon and Carbon–Nitrogen vibrations. The region is generally referred to as the Fingerprint region. Significant bands below 650 cm^{-1} usually arise from inorganic groups, metal-organic groups or lattice vibrations. Tables 1.1–1.5 show the frequency ranges of

Table 1.2. Single vibration and group frequencies and an indication of possible intensities of peaks commonly identified in Raman scattering

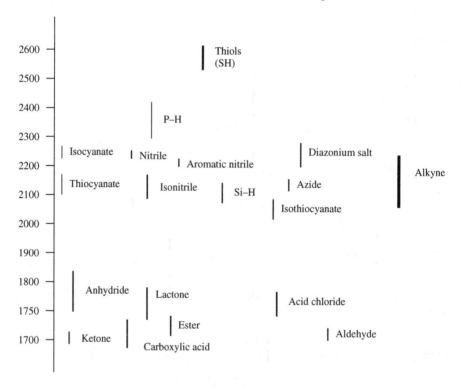

Table 1.3. Single vibration and group frequencies and an indication of possible intensities of peaks commonly identified in Raman scattering

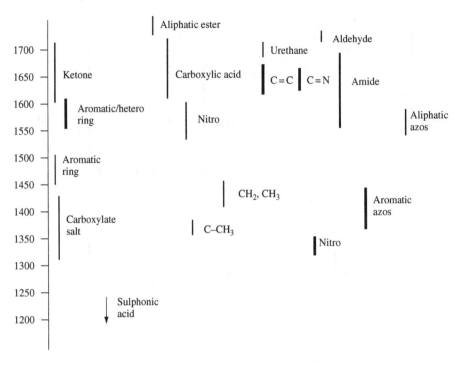

many of the vibrations which give rise to strong bands in either Raman or infrared spectroscopy. The ranges are approximate for the groups in most structures but some groups in unusual structures may give bands outside these ranges. The thickness of the line indicates relative strength. These tables enable a beginning to be made on the assignment of specific bands. A more difficult problem is in estimating the relative intensities of the bands. Earlier in this chapter, we showed that there are reasons why in some circumstances bands which are strong in the infrared spectrum are not strong in the Raman spectrum. However, this cannot be taken as an absolute rule although it is the normal behaviour. Thus, the bands that we would expect to be strong in Raman scattering are the more symmetric bands in the spectrum.

This approach is often used in vibrational spectroscopy. However, to assign specific peaks in the spectrum to specific vibrations, modern laboratories use libraries in which complete spectra are stored electronically. Most spectrometers have software to obtain a computer-generated analysis of the similarities and differences with standards so that specific substances can be identified positively and easily. In other areas, the initial assignment is confirmed by

Table 1.4. Single vibration and group frequencies and an indication of possible intensities of peaks commonly identified in Raman scattering

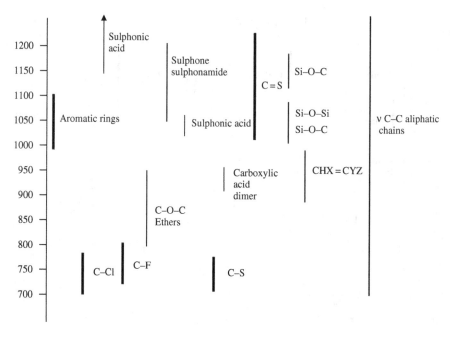

DFT calculations, where the great advantages are a more accurate assessment of the nature of the vibrations and hence of molecular structure.

Predicting the principal infrared absorption bands for small molecules is relatively simple as shown above, but for large molecules, the number of bands possible is very large. Fortunately, many of these bands overlap and what is observed at room temperature are broad envelopes with recognizable shapes in some energy regions and sharp bands due to specific bonds such as $-C{=}O$ in some others. Since some vibrations arise from groups of atoms such as the atoms in a carbon chain or from rings linked by bonds of approximately the same energy, the number of peaks and their energies are linked to the overall shape of the molecule. These are called fingerprint bands and the pattern of these bands can help identify a specific molecule *in situ* in a sample. However, for more complex systems, much time can be spent in the assignment of these bands to the bending, stretching or deformation modes but unless the molecule studied is one of a well-understood set such as an alkane chain of a specific length, this more in-depth analysis does not provide much additional help in the majority of first attempts to identify specific materials from the spectrum.

Table 1.5. Single vibration and group frequencies and an indication of possible intensities of peaks commonly identified in Raman scattering

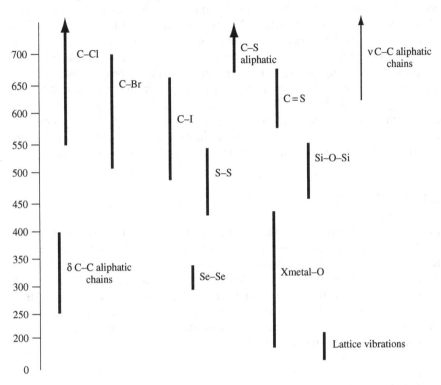

Raman spectra are usually somewhat simpler. The most environmentally sensitive bands, e.g. OH and NH, are broad and weak and the backbone structural bands are strong and sharp. The extent of this difference can be illustrated from the fact that water can be used as a solvent to obtain the Raman spectra of organic molecules. This indicates the relative strength of bands in the organic molecule compared to the weakness of hydrogen bonded species such as the OH bands in water. It is this greater selectivity which leads to the simplicity of Raman spectra compared to infrared spectra. Thus, the Raman spectra of quite large molecules show clear bands. In Figure 1.4 the infrared spectrum is complex and has a strong band just above $1600 \, \text{cm}^{-1}$ from the carbonyl group due to the C=O vibration. The strong bands in the Raman spectrum are largely due to the aromatic group. The band at $2900 \, \text{cm}^{-1}$ due to the CH_2 group is hidden under the strong OH bands in the infrared spectrum but can be clearly seen in the Raman spectrum.

The above information makes it possible to start assigning and interpreting Raman spectra. If possible it is always good to run an infrared spectrum for comparison. The phrase 'interpretation of Raman spectra' is used in many different ways. The spectrum of a molecule can be the subject of a full mathematical interpretation in which every band is carefully assigned or of a cursory look to produce the interpretation 'Yes that is toluene'. However, to be able to carry out a complete, correct and relevant interpretation, the total Raman experiment must be considered. Raman spectroscopists have to make a number of choices in deciding how to examine a sample and the type of answer required may ultimately determine these choices. The simplicity and flexibility of Raman scattering are considerable advantages but if care is not taken in making the correct choices, poor or spurious results can be obtained. Chapter 2 describes the choices and provides the background information to enable the recording and interpretation of Raman scattering in a reliable and secure manner.

1.4   SUMMARY

In this chapter we have attempted to introduce the reader to the basic principles of Raman spectroscopy without going into the theory and details of practice too deeply, with a view to encouraging further interest. Chapter 2 outlines the practical choices to be made in carrying out the Raman experiment in full. Later chapters give the theoretical background required for full analysis of spectra, a guide to ways in which Raman spectroscopy has been successfully employed, and lead to the more sophisticated but less common techniques available to the Raman spectroscopist.

REFERENCES

1. A. Smekal, *Naturwissenschaften*, **43**, 873 (1923).
2. C.V. Raman and K.S. Krishnan, *Nature*, **121**, 501 (1928).

BIBLIOGRAPHY

We have provided a general bibliography. Listed here are a number of publications which the authors have found useful for reference, for theoretical aspects of the spectroscopy and for aids in interpretation.

J.R. Ferraro and K. Nakamoto, *Introductory Raman Spectroscopy*, Academic Press, San Diego, 1994.

P. Hendra, C. Jones and G. Warnes, *FT Raman Spectroscopy*, Ellis Horwood Ltd, Chichester, 1991.

A. Fadini and F.-M. Schnepel, *Vibrational Spectroscopy: Methods and Applications*, Ellis Horwood Ltd, Chichester, 1989.

N.B. Colthrup, L.H. Daly and S.E. Wiberley, *Introduction to Infrared and Raman Spectroscopy*. 3rd Edition, Academic Press, San Diego, 1990.

D. Lin-Vien, N.B. Colthrup, W.G. Fateley and J.G. Grasselli, *The Handbook of Infrared and Raman Characteristic Frequencies of Organic Molecules*, John Wiley & Sons, New York, 1991.

I.A. Degen, *Tables of Characteristic Group frequencies for the Interpretation of Infrared and Raman Spectra*, Acolyte Publications, Harrow, UK, 1997.

D.M. Adams, *Metal – Ligands and Related Vibrations*, Edward Arnold Ltd, London, 1967.

Chapter 2

The Raman Experiment – Raman Instrumentation, Sample Presentation, Data Handling and Practical Aspects of Interpretation

2.1 INTRODUCTION

The Raman spectroscopist has to make a number of choices in deciding how to examine a sample and the choices made are ultimately determined by the availability of equipment and by the type of answer required. Should the excitation source be in the UV, visible or near-infrared (NIR) frequency region? Should the detection system consist of a dispersive monochrometer with a charge coupled device (CCD) detector or an Fourier transform (FT) inter-ferometer with an indium gallium arsenide (InGaAs) detector? Are suitable accessories available to allow the sample to be studied efficiently? How should the sample be presented to the instrument and how can photodegradation and fluorescence be avoided? How do these choices affect the answer? This chapter describes the common types of spectrometer which are used, the accessories available for these instruments, the way in which the sample is presented to the instrument and the way to use data manipulation effectively. The intention is to give guidance in the thought processes required to answer the above questions and others which are essential for a Raman determination to be carried out with confidence.

Modern Raman Spectroscopy – A Practical Approach W.E. Smith and G. Dent
© 2005 John Wiley & Sons, Ltd   ISBNs: 0-471-49668-5 (HB); 0-471-49794-0 (PB)

2.2   CHOICE OF INSTRUMENT

In Chapter 3, when the theory is developed, it will be shown that the intensity of the scattering is related to the power of the laser used to excite the scattering, the square of the polarizability of the molecule analysed and the fourth power of the frequency chosen for the exciting laser. Thus, there is one molecular property, the polarizability, from which the molecular information will be derived and there are two instrumentation parameters which can be chosen by the spectroscopist. This choice is not straightforward. For example, since the scattering depends on the fourth power of the frequency, the obvious way of improving Raman sensitivity is to use the highest frequency possible, which would usually mean working in the UV region. UV excitation also has the advantage that there is less fluorescence than with visible excitation. However, many compounds absorb UV radiation. This and the high energy of the photons in this region means that there is a high risk of sample degradation through burning. It also means that the spectra may be rather different from normal Raman spectra due to resonance with any electronic transition which may cause absorption.This changes the relative intensities of the bands (see Chapter 4 for an explanation of resonance). Additionally, the lasers used can be quite expensive, there is a particular problem with safety since the beam is invisible and the quality of the optics required in the UV is very high. However, the rapid development of optical devices including laser diodes which work in the blue or UV, the unique information that can be obtained in this region and the improved sampling methods now available suggest that UV Raman scattering will be used more widely in the future. Further detail on the employment of UV Raman spectroscopy is given in Chapter 7.

Currently, most laboratories choose either a dispersive or an FT spectrometer. The former employs a visible laser for excitation, a dispersive spectrometer and a CCD for detection. The latter employs an NIR laser for excitation and an interferometer-based system which requires an FT program to produce the spectrum. Both types of instrument have their advantages and disadvantages. The choice for a particular laboratory very largely depends on the type of analysis to be carried out and the materials which the instrument is expected to examine.

2.3   VISIBLE EXCITATION

The most common choice is a visible laser for excitation. These laws are readily available and can be quite compact. Since peaks due to Raman scattering are sharper than the peaks usually detected in the visible region by absorbance and emission, and are measured as a shift from the energy of the excitation source, a monochromatic source is required if quality data is to be obtained. As a result, good quality lasers have to be employed. However, in a modern instrument, weak and broad but recognizable spectra can be obtained even with low-powered, low-cost lasers. Raman scattering using a Raman microscope with a laser pointer has been recorded by the authors.

There are two basic geometries used in collecting Raman scattering: 90° scattering and 180° scattering (Figure 2.1). Both are effective. In 90° scattering, the laser beam is passed through the sample, say in a 1 cm cuvette, and the scattered light is collected at 90° by placing a lens in a suitable position. This light is then imaged onto the entrance slit of the Raman spectrometer. Since the light is scattered as a sphere, the larger the cone of light which can be collected the better. Consequently quite large lenses, or lenses with short focal lengths, are used to cover the largest practicable angle. It has to be remembered that this is not the only consideration. It is also necessary to use the monochromator efficiently and to image the collected light efficiently onto the detector. As a result, the collection lenses have to be matched to the collection optics for efficient performance.

In the 180° system, the laser is delivered through the collection lens and the scattered light collected back through it. In the arrangement shown below, a small mirror is placed in front of the collections lens to achieve this. This is the common arrangement in systems which use a microscope to collect the light. Sometimes a mirror system such as a Cassegranian system or a silvered sphere are used but lenses are more common. 'Grazing incidence' in which the laser beam is directed along the surface is sometimes used in special circumstances.

Some years ago, the spectrometers were quite large. To collect Raman scattering effectively it is necessary to remove the much more intense Rayleigh scattering and other light such as specular reflection from the surface of the sample. This light is much more intense than standard Raman scattering and can flood the detector especially if it is intended to detect Raman scattering which is close in energy to the laser frequency (i.e. low energy vibrations).

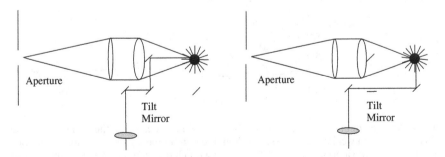

Figure 2.1. 180° (left) and 90° (right) scattering arrangements. The low beam is shown as arriving vertically through a lens and a set of mirrors onto the sample (the black dot) a cone of scattered light is then collected into the spectrometer.

Consequently some device has to be employed to separate the wavelength-shifted Raman scattering from the other light collected. This can be done with two or even three monochromators. The purpose of the first monochromator is mainly to separate the frequency-shifted Raman scattering from the other radiation. The second monochromator increases the dispersion and separates the individual Raman peaks. However, filter technology has improved with the development of effective notch and edge filters. The notch filter, in particular, is widely used. It is designed to absorb all light of the frequency of the incident laser light. Usually a filter which collects most of the light within $200\,\mathrm{cm}^{-1}$ of the excitation frequency is regarded as sufficient (Figure 2.2). The range of frequencies absorbed depends on the quality of the filter and hence the choice is a cost/performance decision the instrument designer has to make. This filter in many instruments can be changed later though other optical re-alignments may also be necessary. Some experiments do require measurements closer to the exciting line. In these cases, the use of monochromators would still be the preferred method for separating the Raman scattering from other light.

The notch filter has a huge advantage in that it is small and efficient. This has very much reduced the size of Raman spectrometers and improved their efficiency. The notch filter type of spectrometer largely dominates the market. In the most

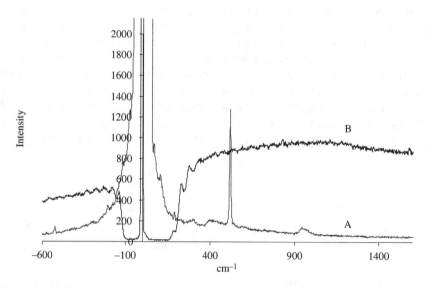

Figure 2.2. Raman spectra taken across the exciting line. In A the Raman band at $520\,\mathrm{cm}^{-1}$ is much weaker than the signal from the non frequency shifted light. The second spectrum **B** is from a poor Raman scatterer with a notch filter in place. Close to the exciting line most scattering is removed by the filter with some laser breakthrough at the laser energy. The features on the edge of the region covered by the notch are artifacts caused by it and not Raman peaks.

common type of instrument, the scattered light is collected through the notch filter and focussed into a monochromator which separates out the different energies of the Raman scattering. The radiation is then focussed onto a CCD. This is a sectored piece of silicon in which each sector is separately addressed to the computer. In this way, it is possible to discriminate each frequency of the scattered light and therefore construct a spectrum of the type shown in Chapter 1. A CCD detector, in absolute terms of sensitivity per photon scattered, is not the most efficient detector in the visible region. However, the fact that all scattering is continually accumulated during the whole exposure of the sample to the exciting radiation more than compensates for this. The choice of a dispersive visible instrument is a good one if flexible sampling, ease of change of excitation wavelength or the use of techniques such as resonance Raman scattering (RR) is a priority.

There are three main disadvantages of this type of equipment. The older systems which used monochromators instead of a notch filter could be tuned to a range of excitation frequencies. To change the excitation frequency with a notch filter system requires that the filter be replaced with a filter of exactly the right frequency for the laser line chosen. Further, filter-based systems do not enable the recording of Raman scattering as close to the exciting line as is possible with the monochromative-based systems scattering at about $10\,cm^{-1}$ from the exciting line has been recorded using additional special filters. The development of simple, reliable, tuneable laser systems may mean that for some specialist Raman spectroscopists, the choice of a double or triple monochromator system may be the more realistic option. However, the simplicity and high state of development of notch filter systems means that these are likely to be the preferred choice for most spectroscopists.

The main disadvantage of using visible excitation is one which is common to all visible spectrometer systems and is a particularly serious one – fluorescence. This is a much bigger problem in the visible region than in either the UV or the NIR. Since Raman scattering is a weak effect, a powerful excitation source is chosen to provide a high power density at the sample. This means that not only can fluorescence occur from the sample being studied but also any minor contaminant which is fluorescent can give large signals. Since fluorescence will occur at energies below that of the excitation, it can be quite intense in the energy region covered by Stokes Raman scattering. Consequently it very often interferes and is probably the main reason visible Raman scattering is not more widely used.

2.3.1 Raman Microscopes

On many modern Raman spectrometers, the sample is simply presented to a microscope which is an integral part of the spectrometer. The microscope has many advantages; for example, it is possible to look at extremely small samples and therefore, despite the fact that Raman scattering is weak, to detect very small amounts of material. Further, it can discriminate against fluorescence from a

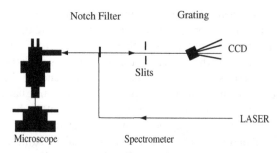

Figure 2.3. Raman spectrometer and microscope, using a visible laser, notch filter, spectrometer and CCD detector.

sample matrix since only the chosen microscopic feature in the sample is irradiated at high power, particularly when the microscope is set up confocally. The effects of this method of sample presentation are described more fully in Section 2.7.

This inherently simple technology has big advantages. Visible laser sources and optics are very good. The coupling of visible spectrometers to a microscope to separate the light collection at the sensing point from the detection system is extremely efficient. Figure 2.3 shows a typical arrangement for a microscope. In this arrangement, the laser is focussed through a pinhole and then collected as an expanded parallel beam. The reason for doing this is to fill the optics of the microscope. There is a plasma filter to remove any specious radiation such as weak emission from lines other than the main exciting line in the laser and any background radiation from the laser. The radiation is then arranged to hit a notch filter. These are interference filters, which work well when the beam is perpendicular to the plane of the filter. At the angle shown in the diagram, the laser radiation contacts the filter so that the light is entirely reflected into the microscope. Once the scattered radiation is collected from the microscope back through the same optics, the beam is incident on the filter at the ideal angle for transmission of the scattered radiation. This light is then passed into the monochromator as shown and onto the CCD detector. Raman spectrometers also have polarizing optics before and after the sample, which is dealt with in Chapter 3.

Another common choice of collection system, owing to the flexibility of fibre optics, is a small probe, designed to excite the sample, coupled to a remote monochromator and detector to collect the scattering. This system is very flexible. There are many uses for which this may be preferred; these are discussed in Section 2.6.5.

2.3.2   Fibre Optic Coupling and Wave Guides

In many applications the use of fibre optics to separate the sampling head from the spectrometer can be a big advantage (Figure 2.4). For example Raman

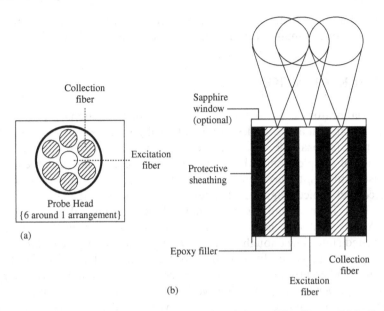

Collection
fiber

Excitation
fiber

Probe Head
{6 around 1 arrangement}

(a)

Sapphire
window
(optional)

Protective
sheathing

Epoxy filler

Collection
fiber

Excitation
fiber

(b)

Figure 2.4. Fiber optic probe end. (Reproduced from J.B. Slater, J.M. Tedesco, R.C. Fairchild and I.R. Lewis, 'Raman spectrometry and its adaptation to the industrial environment' in: *Handbook of Raman Spectroscopy*, Ch. 3, I.R. Lewis and H.G.M. Edwards (eds), Marcel Dekker, New York, 2001, pp. 41–144.)

spectroscopy can be used for on-line analysis on a chemical plant where access can be difficult and the environment not suitable for spectroscopy perhaps because the plant is open to the elements, there is dust from delivery lorries or there simply is no space. However, with fibre optics, only the bead used can be exposed with the expensive spectrometer housed elsewhere. Further, with portable equipment now becoming common and although not necessary for all portable devices, it can be convenient to have a spectrometer which can be held in one hand and a simple small probe in the other. It is also possible to armour these probes to prevent damage. Thus, the use of fibre optics has extended the utility of Raman spectroscopy considerably. One problem that arises is that while passing down the fibre optic, the laser light excites Raman scattering from the fibre optic material itself. This can act as an interference, particularly if the scattering from the sample is weak and a large length of fibre optic cable is used. The main problem comes from Raman scattering produced by the laser beam which can travel in both directions in the fibre. One way of overcoming the problem is to use a multi-mode cable in which the laser is launched down some fibres on the outside and collected through a single central fibre. Usually, the notch filter is placed as close to the collection point as is possible since this will cut out any reflected light or Rayleigh scattering before the Raman scattered light is collected down the fibre. With this arrangement, fibre optic coupling of a sampling head to a Raman spectrometer is an extremely effective way of collecting Raman scattering with big advantages in flexibility.

2.4 NIR EXCITATION

In some laboratories, the overriding criterion for the purchase of a Raman system is that as many samples as possible will give a Raman spectrum. This will be true, for example, in an industrial laboratory where the nature of the next batch of samples cannot be known or controlled. In this case, the use of a NIR laser and an interferometer with detection using an FT program may well be the correct choice (Figure 2.5). The main advantage of the system is that it uses a NIR laser, usually a neodymium-doped yttrium aluminium garnet (Nd^{3+}:YAG) solid state laser emitting at 1064 nm. As a result, few molecules have excited states low enough in energy to give fluorescence. This largely, but not completely, gets round the problem of fluorescence. However, the Raman scattering is inherently weaker because the energy of the radiation is lower and the fourth power law applies. However, since the exciting radiation does not absorb into most samples as efficiently as visible radiation, the laser powers useable are relatively high (up to 2 watts). Further the interferometer detection system, which is essentially the FT-based system used in infrared spectrometers, is very sensitive. The detectors normally used at room temperature are InGaAs detectors which can be cooled to liquid nitrogen temperatures for a slight increase in sensitivity. This type of instrument can also be coupled to a microscope but

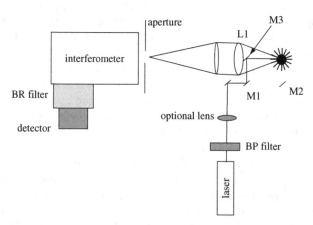

Figure 2.5. NIR FT instrument schematic. The arrangement for collecting the light is very similar to that for a dispersive spectrometer (see Figure 2.1). (Reproduced from Richard L. McCreery, *Raman Spectroscopy for Chemical Analysis*, John Wiley & Sons, Inc., New York, 2000.)

the high power density at the focal point gives an increased tendency to thermal degradation. Very occasionally a sample will fluoresce, but for most samples this is not a problem. Thus, this instrument is closer to an infrared system in that it can record Raman scattering from a wide range of materials present in different states. Compared with infrared absorption spectrometers, the system has its own unique advantages. It is non-contact and samples require little or no preparation. For example, the stringency of the optical system is reduced compared to that of visible systems. Thus, although it is often quite simple to use a visible system to look at a sample in a bottle, this is easier in the NIR because dispersion from non-ideal surfaces is less important and there is less chance of fluorescence from containers.

A comparison of the visible and NIR/FT instruments is made more complex by the fact that the wavelength range of the visible instruments has been extended by using NIR lasers with excitation lines in the 790–850 nm wavelength region. The major problem for the manufacturers of visible source instruments is that the CCD chips lack sensitivity at wavelengths above 1000 nm. This means that lasers that operate at 790 nm or 850 nm are effective but are also very close to the end of the detector range. This can lead to a drop in sensitivity for higher frequency peaks. However these systems increase the range of samples for which Raman scattering can be measured effectively using visible excitation without fluorescence interference.

The range of choice of Raman spectrometers is ever increasing and the size and cost are dropping. A fascinating new development is the production of monochromators with relatively simple CCD detectors which are approximately $4'' \times 2'' \times 2''$ and which use fibre optic coupled sensing. This makes truly portable Raman spectroscopy a much more practical proposition especially since these instruments tend to use simpler CCD detectors which are of lower price than the lab-based instruments.

2.5 RAMAN SAMPLE PREPARATION AND HANDLING

Raman spectroscopy, as a scattering technique, is well known for the minimum of sample handling and preparation that is required. Hendra's rubber duck [1] is a typical example. A small, children's duck thought to be made of rubber was placed directly in the spectrometer beam. Almost immediately a Raman spectrum was recorded of polypropylene! Whilst a large range of homogeneous materials can be examined this way, many samples require some form of preparation and/or mounting in a spectrometer. Compared with infrared spectroscopy, fewer accessories are commercially available though many can be used for both techniques. Typical Raman accessories are powder sample holders, cuvette holders, small liquid sample holders (cf. NMR sample tubes) and clamps for irregularly shaped objects. In the past few years there has also been an increase in the development of specialist cells for rotating solids, vapour cells, reaction cells and variable temperature or pressure cells. In the following section several ways of handling and mounting samples are described, with some advantages, disadvantages and precautions. A review article by Bowie *et al.* [2] has highlighted some of the effects on FT Raman spectra which can originate from the sample. This section gives some examples of how to overcome the more common effects.

Many organic, and inorganic, materials are suitable for Raman spectro-scopic analysis. These can be solids, liquids, polymers or vapours. The majority of bulk, industrial laboratory samples are powders or liquids and can be examined directly by Raman spectroscopy at room temperature. Accessories for examination of materials by Raman spectroscopy are available across a wide range of temperature and physical forms. Sample presentation is rarely an issue in Raman spectroscopy of bulk samples. Many materials can be mounted directly in the beam as neat powders, polymer films, etc. The authors have examined liquids and powders presented in glass containers from capillary tubes, through vials, to 500 ml brown bottles. Samples have also been examined in polymer containers. Raman spectroscopy is less demanding of beam pos-ition, for qualitative work with bulk samples, as the radiation is scattered. However, the sample can be optimized in the beam, particularly necessary for quantitative studies, but the collection solid angle has to be considered. On some occasions the angle of the sample to the scattered beam, i.e. 90° or 180°, will lead to orientation effects. Crystalline samples should be considered from this point of view. Rotating the sample in the beam can average out these effects. Particle size effects have also been reported. The largest problems with samples for Raman spectroscopy occur from fluorescence or burning. Fluores-cence arising from an impurity can, in some cases, be burnt out by leaving the sample in the beam for a few minutes or overnight. This works because there is specific absorption of the light into the fluorophore so that it is preferentially degraded. However, particularly with coloured samples, absorption by the sample can cause degradation of the sample itself. Again this can be reduced by rotating the sample. The speed of rotation has to be kept $<50\,Hz$ in FT Raman spectrometers or beats will be seen across the spectrum. An alternative way to reduce the burning effects is to disperse the sample in other media without a Raman spectrum such as KBr or KCl.

With this type of analysis, one must consider the differing Raman scattering intensities of the analyte and the surrounding matrix and also the possibility of contamination. This is particularly important in Raman scattering since spec-tral intensities can vary quite widely from one substance to another and if the impurity has a strong spectra (or is resonant, Chapter 4) then that spectra can dominate or be an appreciable factor in the final spectrum. There are a number of examples in the literature where this simple precaution has been ignored and important conclusions drawn from data which subsequently has been shown to have arisen from a contaminant. With samples that are in a matrix, e.g. a container or a solution, the relative scattering intensities of the container and the sample need to be considered before the spectra are recorded or interpretation is attempted. An empty polythene bottle placed in the beam will show bands due to polythene. Fill the bottle with sulphur and only the sulphur bands will be observed as the polythene is a much weaker Raman scatterer. Water is a strong absorber of infrared radiation, as is glass. Both are weak scatterers in Raman spectroscopy, which makes the technique particularly suitable for samples in aqueous solutions and/or in glass containers. Glass and water do have their own spectra but only need to be considered with weak solutions.

Small samples may have to be examined with a microscope or microprobe but this means that the beam diameter reduces very significantly and is often much smaller than the total size of the sample. The focal point will then determine which part of the sample is being analysed. This means that it is important to check the homogeneity of the sample by taking a number of measurements across it. It becomes particularly important when larger samples are used with the microscope simply for convenience. The relative refractive indices of the sample and matrix may also have an effect. This is particularly important when attempting confocal Raman microscopy and will be dealt with in Section 2.7.

2.5.1 Raman Sample Handling

Many powders and liquids can be examined by placing the container in which they are supplied directly in the beam. For example, if the sample is supplied in a bottle or vial and the sample gives a strong spectrum, the shape and colour of the vial or bottle is of little importance. Samples have been examined successfully, by the authors, in brown and plastic bottles as well as clear vials. The only constraints are that the outside of the container be clean, free from fingerprints, which cause fluorescence, and the labels do not obscure the sample. If the samples are weak Raman scatterers, then the spectra of the containers can interfere with the spectra of the samples. Not all samples can be examined directly, due to a weak signal, burning or fluorescence. A number of techniques have been developed which reduce some effects and enhance the spectra. Further, plastic cuvettes and microtiter plates can be used with both visible and FT systems. It is important that the laser is focussed into the sample and away from the walls of the cuvette or the sides and foot of the microtiter plate. When this is done, the significantly higher power density of the sample mitigates any interference from the cuvettes or microtiter plates. However, if the sample is focussed onto these materials, excellent spectra can often be obtained from the polymer. Thus, although such systems can conveniently be used and for reasons of cost where repeated measurements are required they may well be the preferred technique, it is essential that the Raman spectroscopist is aware of the possible dangers of the method.

Neat powders with weak signals can be mounted in loosely filled containers or in a compacted solids holder. The authors used the latter technique successfully with a crystalline, low density fungicide which was moved away from the bottle wall by the laser beam power, but gave a strong spectrum 'fixed' in the holder. However, with samples that are crystalline, orientation effects can change the spectra as can the particle size of powders. Using inorganic material it has been shown that the Raman intensity increases as the particle size decreases [3–5]. The theoretical dependence has been described by Schrader and Bergmann [6]. Experimental work has shown that a general fit can be obtained. However, if a sample is dispersed in a matrix, e.g. filler in a polymer or paint resin, droplets in an emulsion, then a sudden in rapid reduction Raman signal can occur at particular sizes below the wavelength of illumination. An example of this is titanium dioxide which gives characteristic Raman spectra in the bulk solid state but gives weak or no spectra when dispersed as a filler in polythene.

Samples of neat powders which are too small to fill the beam or which burn in a bottle can be prepared as a halide disk in a similar way for infrared examination. Strong spectra have been recorded at high laser powers (1400 mw), without burning, by this method. With samples that strongly absorb and burn, very early conventional dispersive Raman instruments employed a sample-spinning device to constantly refresh the sample in the beam. These can cause 'beats' in an FT spectrum, unless the rotation is very slow (<50 Hz). Accessories are now available which will turn samples at this speed [7]. Very sensitive or strongly absorbing samples can burn at very low levels of laser power. The preparation of samples in the same way as hydrocarbon oil mulls (Figure 2.6) for infrared, between salt

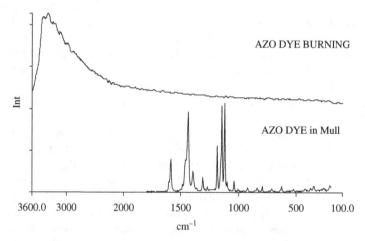

Figure 2.6. Neat sample burning vs. mull spectrum. (Reproduced from J.M. Chalmers and G. Dent, in: *Industrial Analysis with Vibrational Spectroscopy*, Royal Society of Chemistry, London, 1997.)

flats, give good, strong spectra [8]. Even some black samples can be examined by NIR FT Raman at laser powers of 1400 mw. This preparation technique also preserves physical form for polymorphism studies.

A study of the various diluents concluded that KCl was often the best diluent [9]. The process of forming the disks requires pressure and can cause changes in the sample; consequently the making of disks has to be avoided if changes such as polymorphism are suspected or are required to be studied. As already stated, spinning the sample, whilst successful with visible laser sources, has to be kept below 60 Hz with NIR FT Raman spectrometers. The other major difficulty encountered in visible spectrometers is fluorescence. The Raman effect is relatively weak with only ~1 in 10^4 photons interacting with a molecule exhibiting the effect. Fluorescence is generated by very similar interactions with the molecule but is much stronger. Raman spectra can be totally dominated by the broadband fluorescence. Visible laser spectrometers can be used to attempt to burn out the fluorescence by leaving the sample under the beam for some time before measuring the scattering. This may take a few seconds or several hours, or may not occur at all. Moving to higher wavelengths of excitation can significantly reduce fluorescence. Visible lasers with an emitting wavelength of 765 nm have been developed with this advantage in mind. Fluorescence is very much reduced by operating FT Raman spectrometers, with laser sources in the near-infrared at 1064 nm. Although fluorescence is much reduced at 1064 nm, there are cases where even at this wavelength excitation can cause this problem.

Copper phthalocyanine, discussed previously, is unusual in that the visible Raman spectrum shows some background fluorescence (Figure 2.7); the fluorescence increases with exciting frequencies towards the infrared. Early work by the authors showed that blue, green, red, yellow, some brown and even black samples could be examined by FT Raman spectroscopy, but blues and greens based on CuPc would still fluoresce. However, this gives a very characteristic Raman spectrum of many copper phthalocyanine-based materials which show broad fluorescence across the spectrum at 1064 nm excitation.

On increasing the exciting wavelength to 1339 nm, the fluorescence is much reduced and bands due to copper phthalocyanine spectrum reappear. It has been suggested [10] that this strange phenomenon is due to transition metals being present in the phthalocyanine ring. However, this is limited mainly to copper as shown by the spectra, recorded with 1064 nm excitation, of various other metal-substituted phthalocyanines, including metal-free phthalocyanine [11], shown in Figures 2.8 and 2.9. The spectrum of CuPc shown in this figure was obtained by making a very dilute disk (1:1000) in powdered silver reducing fluorescence.

Colour is no guide as to whether a sample will fluoresce. Clear, water white crystals have been observed to cause fluorescence at all illuminating wavelengths. The sensitivity of the Raman spectrum is directly proportional to the exciting wavelength. Spectra obtained at 1064 nm excitation will be less

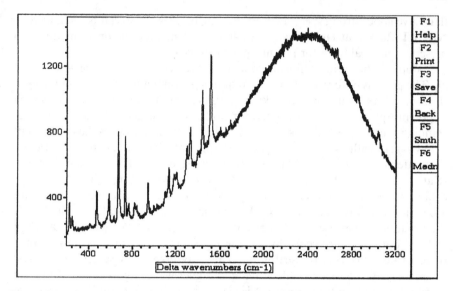

Figure 2.7. Raman spectrum of copper pthalocyanine with 632 nm excitation. (Reproduced from J. Chalmers and P. Griffiths (eds), *Handbook of Vibrational Spectroscopy*, vol. 4, John Wiley & Sons, Inc., New York, 2001, pp. 2593–2600.)

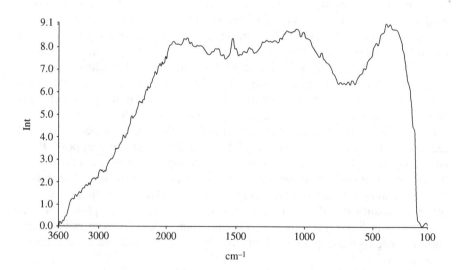

Figure 2.8. Raman spectrum of copper pthalocyanine with 1064 nm excitation. (Reproduced from NIR FT Raman examination phthalocyanines at 1064 nm, G. Dent and F. Farrell, *Spectrochim. Acta*, 1997, 53A, 1, 21 © 1997 by kind permission of Elsevier Science-NL, Sara Burgerhartstraat 25, 1055 KV Amsterdam, The Netherlands.)

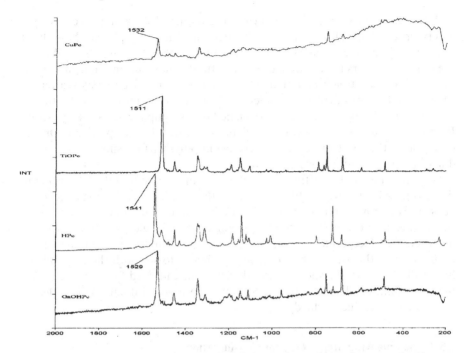

Figure 2.9. NIR FT Raman spectra of phthalocyanines with various metals. (Reproduced from NIR FT Raman examination phthalocyanines at 1064 nm, G. Dent and F. Farrell, *Spectrochim. Acta*, 1997, 53A, 1, 21 © 1997 by kind permission of Elsevier Science-NL, Sara Burgerhartstraat 25, 1055 KV Amsterdam, The Netherlands.)

sensitive than spectra recorded at 765 nm. The fast multiscanning of FT spectrometers helps to mitigate but does not overcome this difference. By moving to lower wavelengths and into the UV, fluorescence is again much reduced. The sensitivity increases but the likelihood of thermal degradation also increases. Even at these wavelengths some solid samples still exhibit fluorescence. Liquids can suffer from fluorescence, but rarely suffer from burning. This is due to their high mobility and hence high heat capacity. If the samples are relatively clear or lightly coloured, spectra can be enhanced by placing the samples in silvered holders. These reflect the signals back through the sample onto the detector. Only small samples can be examined this way or the radiation will be self-absorbed. This effect was demonstrated with tetrahydrofuran (THF) where the $917 \Delta cm^{-1}$ band when excited using 1064 nm has an absolute position of $\sim 8478 \, cm^{-1}$. This is almost at the peak of the NIR absorption band due to the second overtone of the –C–H stretch. Hence, as the path length is varied, the strength of the band will be attenuated due to the self-absorption [12, 13]. Self-absorption has also been observed when using visible lasers in resonance studies.

Polymers of all shapes and sizes can be examined by Raman spectroscopy. Safety spectacles, rolls, thin films, bottles and moulded platens have been examined by this technique. There is a lot of published literature on the Raman spectra of polymers for identification, structural behaviour and morphological properties (see Section 7.3). Polymers are, in general, relatively weak scatterers. This can either be regarded a problem or be used to advantage. As mentioned earlier, samples can be examined in plastic bottles. Sulphur gives a very strong Raman spectrum with no evidence of the bottle wall in the spectrum. On the other hand, the spectrum of a 2% azo dye in polymer film shows both bands due to both dye and polymer.

In examining a polymer film one recommendation is to fold the film as many times as possible to create a 'thick' layer. In this case any orientation effects will be lost. Sometimes film samples are not big enough to fold. An enhanced, strong spectrum can be recorded by placing a small, single sheet flat across the mirrored back face holder. Spectra of coloured polyethylene terephthalate have been recorded this way with strong enough bands to see both the dye and the film. In Figure 2.10 the spectra of both clear film and dyed film were recorded by this technique. The resulting spectrum from a spectral subtraction clearly shows the bands due to the dye.

2.5.2 Sample Mounting – Optical Considerations

As can be seen from the foregoing descriptions, sample preparation and mounting can be relatively simple and flexible. However, if reproducible spectra or quantitation is required, then the optics of the beam and sample presentation needs to be considered. As already described, most spectrometers

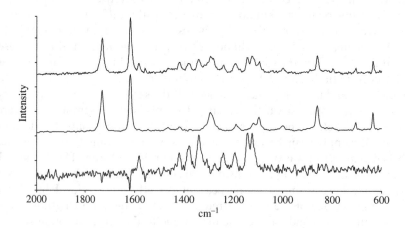

Figure 2.10. Raman spectra of dye in fibre. The top spectrum is from the dyed fibre, the middle one is from the fibre and the foot one is the difference.

collect at a nominal angle of either 90° or 180° to the exciting laser beam. It is the latter aspect which makes the technique so versatile. Also, the radiation is not emitted from a point source. One of the factors which determines the strength of the spectrum is the number of molecules in the sampling volume created by the focussed beam. The sampling volume is the volume of sample irradiated with high power density. It can be a complex state (see Figure 2.11). It can be considered as a cylinder calculated by choosing a power density below which appreciable Raman scattering is not expected (Figure 2.11). The depth of this cylinder is then the distance between the converging beam above the sample with that power density and the diverging beam below the sample with the same power density. The diameter is the diameter of the beam at those points. Clearly, there is no such cut-off point and some scattering may be expected from above and below these lines. The amount will be determined by whether or not the Raman microscope has been set up to be confocal. It is not normally regarded as worthwhile to model the more complex geometric shapes obtained in practice. One good reason for this is that refraction between different fluids and gases causes dispersion of the beam and this is not taken into account in this calculation.

$D = 4\lambda f / \pi d,$
$L = 16\lambda f^2 / \pi d^2,$
D = diameter of cylinder,
L = length of cylinder,
λ = laser wavelength,
d = diameter of unfocussed laser beam,
f = focal length of focussing lens.

This volume can be used to estimate approximately the number of molecules being interrogated in the system at any one time. In the case of a gas, the

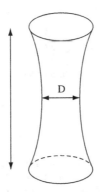

Figure 2.11. Sampling volume for a simple case.

number of molecules will be quite low. As a result, for gas phase measurements, instead of focussing the beam tightly, very long cells are often used. Sensitivity can be increased by using concave mirrors to reflect the exciting and scattered beams back through the sample (Figure 2.12) and in some gas cells, a multiple pass system reflects the beam many times to increase sensitivity. Similar arrangements can be made for liquids but these are usually less necessary due to the higher concentration of molecules in the same volume.

For solids and liquids, as stated earlier, collection should be from as large a cone of scattering as possible. Sensitivity can be improved by using a reflective surface behind the sample tube/vessel. In the case of a modified FT spectrometer, radiation is passed through the Jacquinot stop, which can have a diameter of a few millimetres, or in the case of the visible spectrometer, it is focussed onto the entry slits of the monochromator. To obtain the maximum signal from a homogeneous solid, the surface of the solid should be at, or close to, the focal point. In many cases, acceptable spectra can be obtained from samples not at the focal point. It has been shown that relative band strengths can vary depending on the distance the sample is mounted from this point. Whilst this is not always significant in identification, quantification could be seriously affected [14]. For liquids and gases, the system can be improved further by creating a completely reflecting sphere which gives multiple reflection inside the surface and allows only egress from the sphere through a cone which is collected directly by a lens or is directly focussed into the spectrometer.

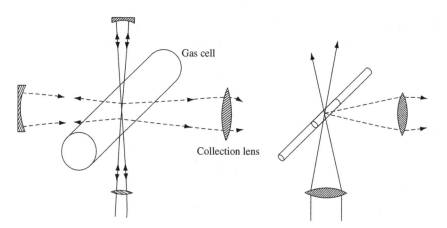

Gas cell

Collection lens

Figure 2.12.   Gas cell and capillary tube both with 90° collection angle. The gas cell shows a double pass system to improve sensitivity. (D. Louden, in: *Laboratory Methods in Vibrational Spectroscopy*, H.A. Willis, J.H. van der Mass and R.J. Miller (eds), John Wiley & Sons, Inc., New York, 1987.)

2.6 SAMPLE MOUNTING ACCESSORIES

2.6.1 Small Fibres, Films, Liquids and Powders

Many samples that cannot be examined directly can easily be mounted at the optimum point by the use of small-diameter glass tubes. NMR sample tubes are often used for liquids, or loosely packed solids, and are easily held in position. Solids can also be held in the open end of the tube, then mounted so that the beam is focussed onto the powder rather than through the glass wall. If the powder in the main part of the tube exhibits thermal degradation (burning), then slowly rotating the tube constantly refreshes the exposed surface. Fibres and thin polymer films can be examined by loosely packing into the tubes or by wrapping round the outside of the tube until a thickness is achieved which will provide a spectrum with a required S/N ratio. Again if burning occurs, the tube can be slowly rotated. Polymers and fibres can also be examined by wrapping them around a glass microscope slide but this would be difficult to rotate. Special cells have been designed [15] to examine fibres and fabrics which involve both sample compression and a back-scattering mirror. The cell also contains a windowless aperture. Previously it was stated that powders which are strongly absorbing can be diluted by KCl, KBr, Nujol, etc. These samples can be mounted in glass tubes in the same way as the neat powders, or prepared as disks and mulls in the same way as they would be for infrared examination.

2.6.2 Variable Temperature and Pressure Cells

Whilst sample heating can be seen as a problem in some cases, there are often requirements to record spectra of samples over a temperature range, both above and below room temperature. This is an area where a wide range of specifically designed cells have been reported to fulfill the requirement for both dispersive and FT spectrometers (Figure 2.13). Cells have been designed to work across a temperature range of –170–950 °C and from high vacuum to 10,000 psig. The difficulty often encountered is in making sealable windows of the optical material – quartz, sapphire and diamond have been used.

Diamond is particularly useful in anvil cells where pressures >1000 atm can be encountered. Whilst samples can be examined over a range of temperatures for reaction rates, morphological changes and degradation studies through Raman spectroscopy can be used in reverse as a temperature probe. By measuring the relative intensities of pairs of bands in the Stokes and anti-Stokes spectra and applying the Maxwell–Boltzman equation (see Chapter 3) the temperature of a sample can be determined.

2.6.3 Special Applications – Thin Films and Catalysts

The examples described so far have been of relatively large bulk samples, and microscopically small samples will be dealt with in Section 2.7. There are

Figure 2.13. AABSPEC variable temperature and pressure cell. (With permission of AABSPEC.)

occasions when less common applications are required. One of these is the study of thin films. Various sampling techniques have been applied, during early modern development of Raman spectroscopy, to obtain spectra at micron or nanometre scales. These were reviewed by Louden [16] for visible spectrometers and include interference enhancement, surface enhancement and attenuated total reflection/total internal reflection.

Figure 2.14 shows an arrangement for examining thin films at a glancing angle with internal reflections along the film enhancing the signal. The scattering nature of Raman spectroscopy at 90° or 180° particularly lends itself to surface studies. SER and SERRS enhancements are described in Chapter 5. However, samples held on solid support materials are open to study, with one of the most common being pyridine, which led to the discovery of the SERS effect [17]. This ability to examine surfaces and interfaces which has led to a number of specifically designed cells for catalysts on electrochemical surfaces at various temperatures and pressures. A typical reaction cell is shown in Figure 2.15.

2.6.4 Flow Through/Reaction Cells and Sample Changers, Automated Mounts

The cells and systems described in this chapter have been on static samples which in many spectroscopists' eyes are a pre-requisite for Raman spectroscopy. Reacting systems, flowing systems or systems which change samples are requirements for the modern spectroscopist. The advantages of Raman

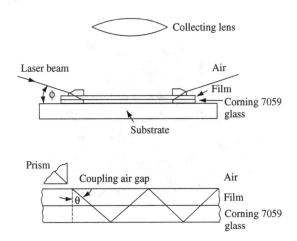

Figure 2.14. Arrangements for measuring thin films. (Reproduced with permission from J.F. Rabolt, R. Santo and J.D. Swalen, *Applied Spectroscopy*, **34**, 517 (1980).)

Figure 2.15. Renishaw cell. (SPD/PN/089 Issue 1.0 June 2003, Product note from the Spectroscopy Products Division.)

spectroscopy, previously described, both in scattering and in being able to probe through glass, provide a great flexibility in approach. Small cells can be simply developed with a glass window for reaction monitoring or simple glass tubes can be used for flow through systems. Automated sample changers have been developed for the semi-continuous examination of pharmaceutical tablets and a combined macro/micro sampling stage has also been developed (Figure 2.16).

In all cases, whilst path length is a minor consideration, the experiment and/ or accessory has to be designed to ensure that the feature of interest matches the beam and the focal point for reproducibility and quantitative considerations

Figure 2.16. Pharmaceutical tablet autochanger. (With permission of Ventacon Ltd.)

(see Section 2.9.3). Other common forms of sample handling are the microtiter plates used in biology. The advantage of these plates is that they allow for multiple analysis simple. Modern Raman spectrometers can give results in a few seconds and the sample can be moved under the beam using software and a sample positioning stage and this is used in mapping is described elsewhere in this book (Section 2.7). Capillaries also are excellent sample holders for Raman scattering. There is some distortion from the walls of some capillaries and this can cause a low background. However, it is possible to seal small samples, single crystals or air-sensitive material inside a capillary and transport it to the spectrometer. No further sample preparation is required. Quartz capillaries and square capillaries can reduce the background significantly.

2.6.5   Fibre Optic and Guided Wave Sensing

This chapter has been primarily concerned with mounting samples in the beam of the spectrometer mainly in the sample compartment. As explained in Section 2.3 the versatility of the technique can be extended by the use of fibre optic probes [18]. We have already described large glass containers being placed in the spectrometer beam. The reverse situation is to take the beam to, or as near as possible to, the sample. For example, the probe can be inserted into a chemical reactor or used to focus on a sample in a container where the environment is hostile or the geometry awkward. By using the transparency of glass, long fibres can be used to examine reactions in vessels from tens to hundreds of metres from the spectrometer on an industrial plant. Some applications are described in Section 6.9. The probes give a high background silicate signal for NIR radiation limiting the distance. Spectra of aspirin tablets have been reported at 50 m using band pass filters [19]. Materials which cannot be introduced into the spectrometer due to their physical size or hazardous nature can have the beam brought to the sample surface. By fixing a lens on the end of the fibre optic probe to focus the beam and act as the 180° collecting window, samples can be examined *in situ*. If a fibre can be pointed at the sample, the spectrum can be measured. Samples which are smaller than the beam diameter can be examined on a microscope by selective use of apertures. The smallest samples examined in this way are theoretically ~1 μm in diameter. This is referred to as the diffraction limit due to the wavelength of the irradiating beam. Applications have been reported [20] where the diffraction limit has apparently been defeated by coupling a fibre optic Raman probe with an atomic force microscope (AFM). In this method, the fibre is metal-coated which prevents egress of the beam from the fibre. It is then heated and drawn out so that an aperture is created at the fibre tip which is typically between 100 and 50 nm in diameter. The light as it passes through the fibre is compressed so that it emerges and rapidly expands from the small aperture. Thus if the tip is placed almost on a surface by the AFM head, the effective excitation area is very small and below the diffraction limit. This process can be reversed where the laser excites the surface externally and the fibre picks up the scattered light. The main problem with the method is that it is inefficient and requires good Raman scatterers to be effective. This is often referred to as scanning near-field optical microscopy (SNOM) which will be discussed in Chapter 7.

Another way of obtaining Raman scattering, which is becoming increasingly popular, is to use waveguiding. In this approach narrow tubes with high-refractive-index tube wall materials are filled with the analyte and the laser beam is launched down the tube so that it is contained in the analyte solution. The signal is then collected at the other end, passed through a notch filter and analysed in a standard Raman spectrometer. The advantage of this arrangement is that there is a very long path length and the laser irradiates the whole sample so that quite dilute solutions can be analysed. The prime requirement is that the sample has a higher refractive index than the sample tube to constrain the illuminating light within the tube and to achieve total internal reflection. This reduces the number of liquids or vapours which can be analysed. Spectra of benzene and weak solutions of sodium carbonate and β-carotene have been recorded [21, 22]. One alternative is to coat the inside of the cell with a high reflective coating such as gold. However this can be limiting in cost and optical sampling arrangements. Another option is to use silver to produce an SERS active surface. Detection limits of less than 10^{-9} mol/L in low refractive index liquids have been reported [23] by using this technique.

2.7 MICROSCOPY

As already mentioned in his chapter Raman spectrometers can have microscopes as an inherent part of the instrument or be easily coupled. The ease of coupling and ability to employ microscopes as sampling accessories comes from the laser sources emitting in the visible region of the spectrum. This means that the scattered Raman radiation can pass efficiently through, and be focussed by, the glass lenses. The major advantage of the microscope is that any sample or part of a sample that can be apertured optically can also have the Raman spectrum recorded. The theoretical spatial resolution is ~ 1 µm. Clearly this will be dependent on the wavelength of the laser source, and the NIR 1064 nm lasers with microscope attached have a theoretical spatial resolution of ~ 5 µm. This is clearly a case of having to decide whether spatial resolution or fluorescence is the bigger issue. The high spatial resolution, and the use of automated stages, enables mapping and imaging experiments to be carried out relatively easily. However there are disadvantages. For example, obtaining a representative spectrum from a film, which may at the microscopic level be inhomogeneous, is difficult. Further, obtaining the spectrum from any significant volume of a solution by detecting a microscopic volume may not be the most effective arrangement. However, if the microscope is an integral part of the instrument, it is possible to buy a small adaptor, which enables collection from a larger volume. These essentially consist of a device which screws into the microscope holder and has a mirror which turns the beam 90° to the microscope direction. A 1 cm cuvette or spinning sample holder can then be mounted on the edge of the microscope stage.

From the optical engineer's point of view, the use of a microscope to detect the scattering has some advantages. A relatively low-powered laser can be used since it will be focussed to give a very small spot giving a high power density at the sample and also a large collection angle. Further, the small excitation volume can be efficiently imaged into a small spectrometer and onto the detector. The microscope can be set up as a simple microscope or can be set up confocally. The advantages of using a microscope have so far focussed only on the X–Y plane. The microscope can also be used to advantage by changing the focus in the Z direction. In the confocal arrangement, the microscope contains a pinhole in its focal plane, which enables only light focussed on the plane containing the sample to be collected efficiently. The pinhole filter stops most other light since it is not focussed sharply in the plane of the pinhole. An alternative system is adopted on some instruments. In this, a slit is placed in the focal plane of the microscope at right angles to the slit of the spectrometer. In this way, the two slits although well separated in the instrument are crossed to create essentially a pinhole. In either case the intention is to discriminate against light, which may arise from anywhere other than from the spot sharply focussed on the sample.

2.7.1 Depth Profiling

With this arrangement, it is possible to obtain some depth profiling of the material. The microscope is focussed at different depths into the sample and the spectra recorded at each depth. From a knowledge of the magnification of the microscope objective used, it is then possible to work out the volume excited and consequently the position in the sample from which the spectra was obtained. Although this would seem in principle relatively simple, there is a considerable problem created by refraction [24, 25] as the beam enters the other material which in general will have a different dielectric constant from that of the air between the sample and the microscope. This can be decreased to some extent by using water immersion and oil immersion objectives, but in general considerable care must be taken when estimating the true depth into the sample. However, given this limitation it is still possible to obtain some information about how a signal changes with depth in the sample. A depth profile taken with a microscope with 632 nm excitation for the polymer polyethylene terephthalate is shown in Figure 2.17.

2.7.2 Imaging and Mapping

The techniques of Raman mapping and imaging arise from the use of microscopy with Raman spectrometers. The CCD devices used as detectors are essentially similar chips to those used in digital cameras and camcorders. They are arranged in an array of pixels each of which can be individually addressed. The difference in Raman scattering is that, since the signal is weak, the background noise is critically important and consequently in most instruments, the sample is cooled using three-stage Peltier cooling or even liquid nitrogen cooling. Some less expensive spectrometers employ standard chips used in cameras either with no cooling or with single-stage cooling. In the normal arrangement, the scattered light is separated into individual frequencies in the monochromator and focussed as a line on the CCD so that each separate frequency can be detected at a different point along the line. An alternative way of collecting Raman scattering is that instead of using a monochromator to split up the different frequencies, a set of filters can be used, in a manner analogous to that used by Raman in the initial experiment. In this arrangement only light of a particular frequency range corresponding to the frequency of one of the major vibrations of the molecule to be detected can pass through to the detector. In this arrangement, there is no monochromator to split up the light and the detector operates exactly like a camera recording an image of the sample focussed under the microscope. The only difference is that only light of the frequency of the Raman active vibration can reach the detector so a Raman image is recorded. This is called imaging. Figure 2.18 shows a combination of effects [26]. The photomicrograph and corresponding Raman image of the $1374 \, cm^{-1}$ band clearly shows a parasite's food vacuole along with the spectra of hemozoin, L-hematin and hematin all acquired using 780 nm excitation. The spectrum of hemozoin is identical to the spectrum of L-hematin at all applied excitation wavelengths. The band enhancement of A_{1g} modes, explained in

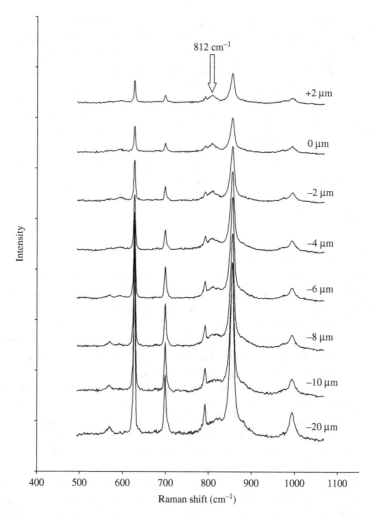

Figure 2.17. Confocal depth profile of 78 μm acrylic latex coated on 200 μm PET coating. The depth resolution was determined to be 4 μm with a ×50 objective at 632.8 nm. Spectra were collected from the coating surface (0 μm) and at 2 μm intervals through the polymer (spectra labelled −2 μm through to −20 μm). The uppermost spectrum was collected when the laser beam was focussed 2 μm above the coating surface, so that the Raman scattering arose from the residual, or defocused, part of the beam. (Reproduced with permission from G.D. Macanally, N.J. Everall, J.M. Chalmers and W.E. Smith, *Applied Spectroscopy*, **57**, 44 (2003).)

Chapter 4, including X4 (1374 cm^{-1}) enables Raman imaging of hemozoin in the food vacuole. This enhancement, resulting from excitonic coupling between linked porphyrin moieties in the extended porphyrin array, enables the investigation of hemozoin within its natural environment.

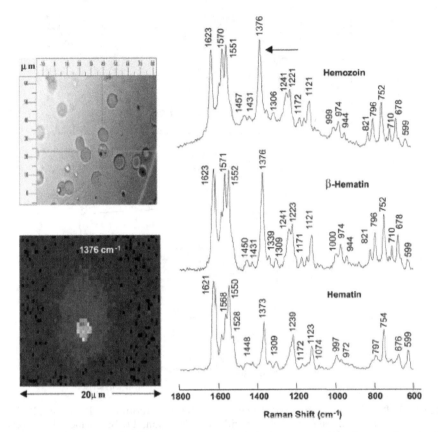

Figure 2.18. Raman imaging. Photomicrograph and corresponding Raman image of the 1374 cm^{-1} band. (Reprinted from B.R. Wood, S.J. Langford, B.M. Cooke, F.K. Glenister, J. Lim and D. McNaughton, *FEBS Letters*, **554**, 247–252 (2003).)

An alternative method is to map the surface. Accurate positional devices are now readily obtainable and using a suitable XYZ device, it is possible to use the standard configuration to take a Raman spectrum from a small area, move the sample so that the next small area is under the microscope and take another spectrum. By doing this repeatedly, spectra from a selected area can be obtained. From this data, any one vibration can be selected and a map of the intensity variation for that vibration plotted. Figure 2.19 shows typical maps. One is a black and white image of the surface with each pixel shown being a point at which a spectrum was taken. The lighter the pixel, the more intense the Raman scattering recorded. The other map shows a 3D representation. These maps were obtained from a surface in which a number of small particles, which give extremely strong Raman signals, had been deposited. The position of the particles can clearly be seen from the peaks shown in the map.

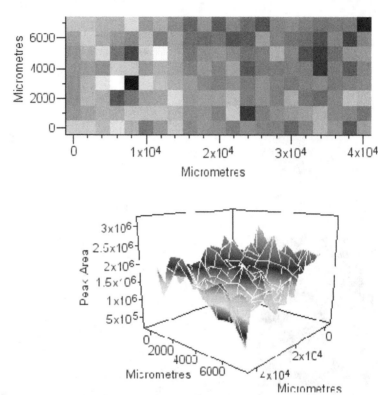

Figure 2.19. Raman maps – pixel map (top); 3D map (bottom). (Reproduced with permission from A. McCabe, W.E. Smith, G. Thomson, D. Batchelder, R. Lacey, G. Ashcroft and B. Fulger, *Applied Spectroscopy*, **56**, 820 (2002).)

Imaging has the advantage that it is rapid but the disadvantages that only one particular region of the spectra can be surveyed at one time and the resolution is limited. Mapping has the advantage that the whole spectrum is recorded and stored on the computer. A map of all vibrations observed could be obtained if required. However, it is very slow in practice. Obtaining Raman spectra by either of these techniques but in particular by mapping has large advantages in that the material can be immediately identified from the spectrum information obtained and its distribution in a heterogeneous sample determined. One use of this technique has been to study the distribution of drugs in tablets by mapping the surface. This identifies not only the drug but also other components such as fillers and binders and gives their distribution. Apart from the value of this information in formulation, distribution in a tablet is difficult to counterfeit. The best maps of this type can take days to complete.

2.8 CALIBRATION

So far we have considered the spectrometer components and sample presentation. Before continuing to examine and manipulate the data, a question which should regularly be asked is, 'How do you know that your instrument is working correctly and consistently?' It is a question being asked of industrial spectroscopists more and more, specifically by regulatory authorities. As Raman spectroscopy is growing in industrial use, the questions are being rightly asked by non-scientists. The pharmaceutical industry, in particular, have to register new products before sale to the public. The regulators wish to know that the measurements have been made correctly, on correctly working instruments which will give the 'same' answer today and tomorrow on the 'same' or similar samples. The question is particularly important if quantitative work is carried out. Apart from regulatory requirements, industrial spectroscopists often require quantitative methods to be transferred between instruments. This is a topic in which interest has recently increased particularly in NIST and ASTMS. The search for a simple standard and method of calibration continues but at least one daily check has been published by McCreery [27].

Most of the checks that are carried out are to ensure that the x-axis or wavenumber position is correct. The phrase 'calibration' is often used but that is not possible for most spectroscopists; it can be done only by engineers. The checks that are carried out are better described as performance checks. Barium sulphate has a strong band at $988\,cm^{-1}$, diamond a band at $1364\,cm^{-1}$ and silicon a band at $520\,cm^{-1}$ which are now used by some instrument manufacturers. In addition, indene [28], cyclohexane and sulphur have well-known band positions as measured on dispersive instruments. However calibrating relative peak heights is a rarely mentioned field. For NIR FT Raman spectrometers the situation is worse. Whilst the sulphur spectrum maintains relative band strengths, indene bands vary greatly in relative intensity with laser power (Figure 2.20).

Whilst the spectra appear very similar, the ratio of the 2890 to $1550\,cm^{-1}$ bands does not change linearly with a change in laser power from 10 to 350 mw. A number of compounds with absorption bands in the NIR spectrum above the laser line at 1064 nm show this effect. This appears to particularly affect compounds with aliphatic hydrocarbon groups. The bands with a Raman shift of $\sim 3000\,cm^{-1}$ are actually scattered at a true frequency of $6398\,cm^{-1}$ which is equivalent to 1562 nm. This is very close to the broad NIR aliphatic hydrocarbon overtone bands at approximately 1666 nm, halogenated dienes have been suggested as a possible standard. The vCH bands, and others near to the limits of the detector range, are also frequently strongly attenuated compared to spectra excited with a visible light laser. Bands are likely to be severely attenuated when they occur close to the cut-off edges of the filter used to block the elastically scattered radiation occurring at the exciting line frequency. Recently, standards based on fluorophores have been proposed for spectrometers with visible laser sources.

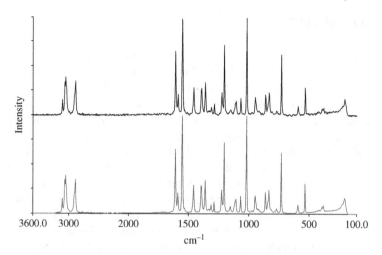

Figure 2.20.   Indene at 1064 nm – 10 mw (top); 350 mw (bottom).

A neon lamp on the beam axis can provide a wavelength calibration standard. Halogenated dienes and cyclohexane have been suggested as possible Raman wavenumber standards [29]. The latter does have a variable spectrum dependent on the wavelength of the laser source used. The spectrum has to be corrected for instrument response as shown in Figure 2.21.

An ASTM standard (ASTM E 1840) has now been established for calibrating the Raman shift axis. Eight common chemicals – 1,4-bis(2-methylstyryl) benezene, naphthalene, sulphur, 50/50(v/v)toluene/acetonitrile, 4-acetamidophenol, benzonitrile, cyclohexane and polystyrene had the Raman spectra recorded by six different laboratories using both dispersive and FT spectrometers. Apart from a few of the values at high and low frequencies, standard deviations of <1 cm^{-1} were reported.

The instrument response is variable across the *x*-axis. Using tungsten lamps is often quoted as the way to measure instrument response. Unfortunately the lamp energy is dependent on temperature and this varies with the lifetime of the bulb. For an accurate calibration, the filament temperature would have to be measured. The use of tungsten lamps and glass filters has been proposed by NIST [30] and adopted by instrument manufacturers [31] to overcome this issue, particularly for transfer of quantitative methods. To calibrate the *y*-axis, a simple, practical calibration standard for instrument response correction, based on the use of luminescent standards (fluorophores), was proposed [32]. These have been developed further by using correction polynomials. These are available via the Internet [33].

Whilst standards are now becoming more available, there is still not a universal, easy to use, sample which calibrates both wavenumber and intensity

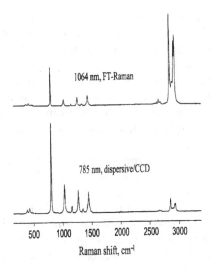

Figure 2.21. Cyclohexane uncorrected for instrument response. (Reproduced from Richard L. McCreery, *Raman Spectroscopy for Chemical Analysis*, John Wiley & Sons, Inc., New York, 2000.)

in a single spectrum. The luminescent standards have to be used in the same sampling geometry as the sample of interest. The wavenumbers position can be affected by several instrument features particularly in FT systems [34].

2.9 DATA HANDLING, MANIPULATION AND QUANTITATION

Having organized the Raman experiment with regard to sample presentation and instrument operation, we need to consider how the data will be generated and manipulated. The latter will depend on the use to which the data is put. As already stated, the phrase 'interpretation of vibrational spectra' is used in many different ways. In a qualitative way the spectrum of a molecule can be the subject of a full mathematical interpretation in which every band is carefully assigned or of a cursory look to produce the interpretation 'Yes that is acetone'. Alternatively the spectra could be employed to monitor or determine composition in a quantitative way. Whichever way the data is used, the manipulation which has occurred during production has to be considered.

2.9.1 Production of Spectra

Raman instruments are single-beam instruments that are operated in the vast majority of cases without the need for a background reference spectrum. The nature of the spectra produced can be affected both by the Raman instrumentation and by the way in which the data is manipulated following collection. However instrumental features such as filters or beam splitters can affect the spectrum. Dispersive instruments may employ more than one filter or monochromator to cover a wide spectral range. At the change over point energy difference can affect the band strengths although this is less of a problem as modern instruments. In FT spectrometers the raw data is not a spectrum but an interferogram. This is computer manipulated before presentation as a spectrum.

The relative strengths of the bands in the $3000\,\mathrm{cm}^{-1}$ region are particularly affected in FT Raman and as discussed above the use of 792 or 850 nm excitation with visible Raman systems can also affect relative intensities. Perhaps the biggest feature is the direction of polarization of the laser beam but this will be discussed in Chapter 3. Background correction of spectra can be carried out with a white light source. In the ideal world this would have a known, invariable temperature. In practice this does vary with time and can cause variations in the background as can a change of filters, etc. These effects are most critical for quantitative measurements rather than qualitative measurements. The effects of apodization and resolution can be seen in the spectra from FT Raman instruments. A major advantage and also a problem with modern instruments is the flexibility of the software used following data capture.

Before digital displays were commonly used, if the spectrum was weak, this was instantly obvious. Now it is easy to produce an apparently strong spectrum by simply changing the intensity scale. This is often carried out automatically by the instrument. If care is not taken to read the Y scale, the information that the spectrum is weak can be missed. A weak spectrum may be due to too little sample, poor preparation, the fact that the 'sample' is loaded with a diluent such as salt or simply that the sample is a poor Raman scatterer. In the last case, the spectrum may be from an impurity due to a strong Raman scatterer in the sample matrix. Further, spectra are often scaled for comparison with each other by overlaying. This is usually carried out by choosing a band common to both spectra. The experienced spectroscopist will look at the noise present on the spectra away from the main peaks to judge the relative intensity at the peaks of the spectra in their original form. However, modern programmes also contain smoothing routines which on many occasions can be quite useful. In this case, the smoothing routine can be used to remove all noise from the spectra and prevent an estimate of intensity.

Spectra can be subjected to manipulation by expansion, smoothing, baseline flattening, spectral subtraction and numerous other software programs, not forgetting the effects of the apodization function in the FT instruments. Once the initial spectrum has been produced, any further software manipulation or enhancement should be approached with great care. In many cases the manipulation can make the spectrum look pretty but will result in the loss of vital information for interpretation. These routines should only be used with an understanding of what may be occurring!

Perhaps the most dangerous feature of spectrum manipulation is overuse of smoothing programmes. They can often be used to make very small features of the spectra into large looking features by selecting the spectral range recorded and smoothing out any noise present in the spectrum to obtain a smooth looking band, which gives the impression it is a major feature. The fact that the band is a minor feature does not make it unimportant. It may well be extremely important. It does however increase the possibility that it arises from an impurity or other spurious cause in the spectrum. Overall however, if used correctly, these modern data handling programmes are extremely effective.

2.9.2 Display of Spectra

Spectral presentation is generally not an issue. Raman spectra are usually presented as just the Stokes spectra with the anti-Stokes spectra omitted. The only inconsistent feature is in the way in which the wavelength scale is displayed, sometimes from high to low wavenumber but often from low to high wavenumber. There are semantic debates as to which is correct. Purists say that all graphical scales should be displayed with the lowest value at the origin. Others say that the wavelength scale, in Raman spectrometers, is a shift and not an absolute measurement. For comparison with infrared spectra, the high to low format is preferred. In this format both infrared and Raman spectra from the same sample can be overlaid and band positions compared. All bands in a Raman spectrum rise from a baseline against a linear scale. This scale varies from instrument to instrument and cannot be easily used for direct quantitative measurements. They can be used for comparative measurements and band ratio quantitation.

❑ SPECTRUM SCALES

Modern instruments almost invariably produce spectra which fill the screen or the page of a printout. Two spectra can be given equal importance if careful observation is not made of the intensity scale. Data systems automatically scale spectra so that the strongest peak is that which stretches to the top of the screen. By ignoring this effect information about the sample can be lost. As previously stated, a weak spectrum can be due to instrument effects, sample mounting, diluents or the sample having a low Raman scatter cross-section. If spectra of samples from different points in a matrix appear to be of the same strength, just because of scale expansion important information can be lost or misinterpreted.

A legitimate way to scale spectra for comparison purposes is normalising. This is usually carried out by adding a standard or choosing a band common to both spectra. The absorbance of the band in the strongest spectrum is fixed and the same band in the comparison spectrum scaled to fit. Other bands in the spectra are scaled but the relative intensities to the main band remain the same.

❒ SPECTRAL ENHANCEMENT/LOSS OF DATA

A number of data handling packages available for manipulating spectra are meant to enhance the appearance of the spectrum, particularly for inclusion in reports or publications. Although many claim to improve, or facilitate, interpretation, if not used with great care the opposite effect can be achieved. Software packages are available which will reduce or remove the slope from all or part of the spectrum. This can be very useful to Raman spectroscopists to remove fluorescence backgrounds, but could lead to wrong assumptions on purity or affect quantitative measurements. If quantitative work is being carried out, another way of correcting for a background slope is to carry out derivative spectroscopy. As can be seen in Figure 2.22 the spectrum of the second derivative has a flat baseline but identifying individual bands can be complex.

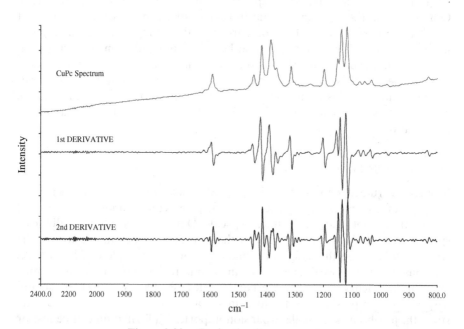

Figure 2.22. Derivative Raman spectra.

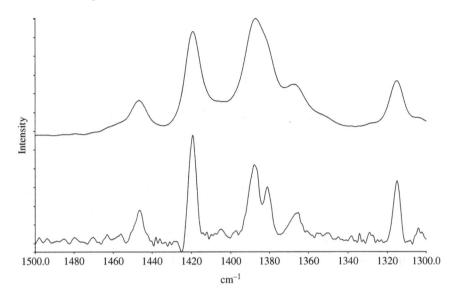

Figure 2.23. Deconvolution of Raman bands from a broad spectrum.

Many data systems contain routines to produce peaks from within a broad envelope as can be seen in Figure 2.23. These are referred to as deconvolution routines. These are often useful for quantitative studies when the number of components is known. Good judgement has to be made as to when to stop if the number of components is unknown. In the latter case information is not lost as much as spurious bands are produced. An example of these routines being used effectively is given in Chapter 4 (Figure 4.9).

The SMOOTH function is one which can be used as a precision tool to enhance a spectrum or a blunt instrument to totally destroy it. If a spectrum has a low signal to noise ratio, the effects of the noise can be reduced with a light smooth to make the bands appear more clearly. A heavy smooth can not only lose shoulders, but remove some bands completely. Figure 2.24 shows a noisy spectrum with several degrees of smooth.

2.9.3 Quantitation

The data handling procedures have concentrated on the effects on the qualitative aspects of the spectrum, but vibrational spectroscopy is also used quantitatively. For most uses of Raman scattering, the essential feature is the ability to detect the spectrum. The frequency values are given with reasonable accuracy. Intensities, on the other hand, are usually treated as relative intensities or described as 'strong', 'medium' and 'weak'. This is sufficient information if what is required is to use the spectrum as a fingerprint for the molecule or

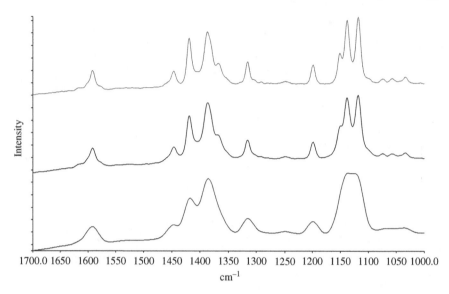

Figure 2.24. Smoothed Raman spectra with the foot spectrum showing the danger of over smoothing.

molecules from which the scattering occurs. On normal infrared laboratory instruments, quantitative analysis is carried out either by measuring the absorbance of a specific band or by ratioing the absorbances of component specific bands. As Raman spectroscopy employs a scattering technique, the absolute band intensity is dependent on a number of factors such as orientation, laser power and other instrumental effects which cannot be reliably reproduced. Until recently absolute band intensity measurements ought to have been avoided. Advances in instrumentation and stability are starting to open up this area of application. Relative strength determinations can be made, in multicomponent samples, by measuring the ratio of intensities of relative bands. However the relative Raman scattering cross-section should be borne in mind for each component. The calibration curves should ideally be constructed from similar samples of known composition. For more than two components, synthetic mixtures do not usually provide adequate calibration curves. Real samples can contain minor components which contribute to band shape and size. Components may also interact with each other and solvents to produce variations in peak position, shape and size. As already mentioned, particle size, self-absorption and depolarization ratios can all affect relative band strengths.

Bands may be measured in various ways. The most common method is peak height. A baseline point must be established which takes account of band shape, neighbouring bands and possible fluoresence. The absolute position of the base point is not critical but the method of determining the point must be consistent in the calibration and test samples. All the spectrometer conditions, sample concentrations, sampling volume and, where relevant, cell window material or reflection angles must be noted, if the measurement is to be repeated or transferred to different instruments. Relative band intensities will vary with different wavelength laser sources as there is $\sim v^4$ dependency on signal intensity with absolute wavenumber. As already mentioned self-absorption can affect band intensities, particularly in NIR Raman. Relative intensities can also be enhanced by resonance as described in Chapters 4 and 5. Temperature may also be critical in some measurements, especially at high laser power or with a strongly absorbing sample which could under go thermal degradation.

Raman spectroscopy carried out in most experiments is not an absolute quantitative method but mainly comparative. However, quantitative information can be obtained from single peaks with the relevant data treatment and careful consideration of the factors which may affect the result. These fall into two categories – those which affect production of the data, being hardware- or sample-related and those which occur during data handling.

☐ QUANTITATION – HARDWARE AND SAMPLING FEATURES

All Raman spectrometers are essentially single beam instruments and consequently any quantitative analysis procedure will be dependent on the stability of the laser and the detector. Further, the same instrumental condition will require to be used on each occasion the analysis is carried out. It is unlikely that this can be done without significant fluctuation due to changes, for example, in laser power with time. As a result, all quantitative measurements using Raman scattering should make use of a calibrant which should be run at the same time and preferably interspersed with the samples used in the quantitative procedure. All the spectrometer conditions, sample concentrations, sampling volume and, where relevant, cell window material or reflection angles must be noted, if the measurement is to be repeated or transferred to different instruments. Temperature may also be critical in some measurements.

One problem is that the flexibility of Raman sampling can mean that reproducibly replacing a sample may be difficult. The use of focussed beams in Raman scattering to obtain higher power means that relatively small volumes of solution are usually interrogated. For example, although a 1 cm cuvette is used, the volume of the sample within the cuvette actually sampled is usually of the order of microlitres. To obtain effective Raman scattering from this volume, the beam must first of all pass through the holder and therefore be refracted by it and then through the media. The scattered radiation then repeats the process on the way back to the detector. The depth at which the sample is focussed can alter the signal, and any mis-alignment of the cell which causes a slight displacement of the laser beam can also affect signal intensity. Thus, it is essential that a stable holder is built which defines the position of the sample in relation to the collection optics. It is also essential that the instrument parameters be set in exactly the same position on each occasion. If this is done, and provided a statistically large enough number of spectra are taken per sample and regular standards are included in each set of measurements, effective quantitation can be obtained. The standards used should span the wavelength range of the peaks measured in the analyte. The use of a silicon standard with a peak at about $550\,cm^{-1}$ with an organic sample for which the main peaks are in the region $1000-1600\,cm^{-1}$ is poor practice. In principle, the instrument should cope with the difference in frequency, but the authors have obtained changes caused on one occasion by a slipping stage on a grating in the instrument and in another with the use of near-infrared radiation with a visible system where the detection of the Raman lines was close to the edge of the range for the detector. This caused differential changes in the relative intensities of the bands depending on frequency.

As with other optical techniques, it is easier to quantify solutions or gases rather than solids where the nature of the solid can have a large effect on the spectra obtained. The use of the micro sampler attached to a microscope, or the use of a system without a microscope, so that a representatively large amount of the solution is detected is also a help. Examples of the use of Raman for quantitation are given in Chapter 5.

Clearly, a double beam approach would be more effective for quantitation. However, many modern UV visible spectrometers designed for quantitative use no longer have the double beam arrangement. One possible approach is to use a cell split into two with one half filled with sample and the other with the standard. The cell is spun so that the standard and sample spectra are recorded regularly over a period of time. The result is then obtained from the average accumulated signal. However, these can be difficult to fill and to use. The reliability of modern machines is sufficiently good not to need to do this. The most important variable to check is the laser power which if not automatically compensated can drift considerably over a day. Recently, instruments which can record quantitative spectra in the standard 96 or 384 well-microtitre plates widely used in biology have become available.

◻ QUANTITATION – DATA HANDLING CONSIDERATIONS

The intensity from which quantitation is to be obtained can be measured in various ways. The most common method is to measure the height of a major peak, but peak area can sometimes be a better measurement. At some point in the spectra, a baseline which shows no Raman scattering must be established; this should take account of band shape and neighbouring bands. The absolute position of the base point is not critical but the method of determining the point must be consistent in the calibration and test samples. As with the qualitative manipulations, there are a large number of quantitative software packages available. Some can be used for composition analyses which attempt a simple least squares fit, through principle component regression (PCR) to partial least squares (PLS) modelling. Besides these, spectral enhancement and band resolution packages are available on many instruments. Simple derivative spectroscopy has already been mentioned, but Fourier domain processing and curve fitting routines, sometimes in complex combinations, can also be employed. All must be employed with an understanding of the applicability of the package used to the problem being studied. Otherwise the result can be due more to the imaginative component than the real.

A number of texts delve deeply into the mathematics of the quantitative aspects of vibrational spectroscopy. We have only highlighted here the features which are of particular note for Raman spectroscopists.

2.10 APPROACH TO QUALITATIVE INTERPRETATION

In Chapter 1 the basic theory of and approach to interpreting a Raman spectrum were set out. In this chapter we have considered various instrument features such as the source wavelength, accessories which give effective sampling and how data production can affect the final spectrum. All these factors should be borne in mind before any attempt is made to interpret a spectrum. In fact some of these parameters may have been specifically chosen to enhance a particular feature of interest. In Raman spectroscopy whole techniques can be devoted to specific enhancement as will be seen with the SERRS effect in Chapter 5. If the basic structure of a molecule is known, then the theory expounded in Chapters 1, 3, 4 and 5 will assist the spectroscopist to make great progress towards gaining chemical, physical and even electronic information about the state of the molecule. Subtle and not so subtle changes in bands can yield extensive specific information about the molecule. Vibrational spectroscopy is often used in attempts to identify unknown materials, to characterise reaction by-products and to follow reactions. Raman spectroscopy in this context is a poor cousin of infrared spectroscopy and to some extent has been oversold. It should be noted that although Raman spectra are often simpler and clearer than infrared spectra, they can be less easy to fingerprint since some groups do not give strong bands and there are far fewer published, recorded reference spectra for direct comparison with unknowns. However, like most tools when used with skill and the correct approach, Raman spectroscopy can be of great assistance in identifying unknown materials or components. To do this successfully the maximum information about the sample should be obtained and borne in mind during the analysis and it is essential to be aware of problems which can lead to an erroneous result.

2.10.1   Factors to Consider in the Interpretation of a Raman Spectrum of an Unknown Sample

In a practical interpretation it is essential that all available information is used and that the possibility of contamination is considered. There are a number of examples in the literature of this simple precaution being ignored and important conclusions drawn on data which subsequently were shown to have arisen from a contaminant. Whilst in both Raman and infrared spectroscopy, interpretation of the spectrum requires knowledge of all the factors which may be affecting the spectrum, Raman spectroscopy has fewer complexities. Sample preparation is often zero with samples being examined as neat solids, liquids or gases, with only a few possible artefacts. Some instrumental effects such as cosmic rays and emission from room lights and in particular strip lights and cathode ray tubes can show up in the spectrum. These appear abnormally strong in weak spectra where scale expansion has been used. Some but not all of these features can be recognized because the bandwidth is narrow but it is essential that thorough checks are made for the presence of these peaks.

It is important not to lose sight of the overall picture. If simple information on the nature of the sample is ignored, answers can be generated which common sense tells us are impossible. The very different intensities of Raman scattering from various vibrations from different molecules in a matrix can easily lead to this sort of wrong interpretation. A polymer bottle may contain sulphur. Polymers are weak scatterers whereas sulphur is a strong scatterer. The fact that the spectrum is dominated by the sulphur peak does not mean that the polymer is largely sulphur. This is a rather trivial example but this mistake is easy to make when two organic molecules are present in a matrix.

Raman spectra are not obviously dependent on the chemical and physical environment of the sample being examined. Whether the molecules are in a gaseous, liquid, solid or polymeric form is not easily apparent from the spectrum, but the physical state does affect the overall strength and band shape. In general, crystalline solids give sharp, strong spectra whilst liquids and vapours tend to have much weaker spectra. Pressure, orientation, crystal size, perfection and polymorphism may affect the spectra, but the changes can be subtle. Raman spectra are however particularly temperature-sensitive. Broad bands in Raman spectra tend to be due to fluorescence, burning, low resolution or weak bands that have been enhanced, e.g. from glass or water. Chemical groups may also respond to hydrogen bonding and pH changes but these changes tend to be shown in peak shifts rather than changes in band shape.

So having recorded the spectrum, we need to develop an approach which will help as much as possible towards solving the problem. By sequentially going through the next steps the chances of making an error in interpretation will be much reduced but success is not guaranteed!

❏ KNOWLEDGE OF THE SAMPLE

A lot of information can be gained by understanding the way in which a sample arrived in the state presented for examination. The analyst should consider the following questions. How was the sample produced? What is known of the reaction scheme? Are there possible side reactions? Could solvents be present? Did work-up conditions introduce impurities? What was the type of equipment the sample came from? (Grease, drum linings, coupling tubes and filter aids can all appear in or as the sample spectrum.)

Solids – Is it 'dry', or a paste? Has it been washed with a solvent or re-crystallized?

Liquids – Are they volatile, are they alkaline, neutral or acidic?

Vapours – What temperatures/pressures are involved?

How pure is the sample thought to be? Is any elemental information available, does the sample contain N, S, or halogens? Could these come from an impurity?

Are there likely polarization, orientation or temperature effects?

The answers to these questions are not always available but these points should be kept in mind if the spectrum does not appear as expected. The spectrum of a sample without any known history or source should be approached with great care. Something is always known even if it is only physical form and colour.

❏ SAMPLE PREPARATION EFFECTS

Handling the sample may affect the resultant spectrum; as mentioned in previous sections, the information required may dictate the sample preparation and/or presentation method. Knowing the method should provide some information about a sample but beware.

- Solids – Is it neat, a halide disk or mull? If it is a mull, mark off any bands from the mulling agent. Is the sample a neat powder, could this produce orientation or particle size effects? If the sample has been diluted because it has a strong colour why say it is a colourless material?
- Is the sample is in a container? Mark off the bands due to the vessel walls, e.g. glass, polythene.
- In cast or polymer films, is there any solvent trapped or encapsulated? Can polymer films have orientation?
- Liquids – Is the sample a pure liquid or a solution? If the latter, mark the solvent bands.
- Microscopy – Are the bands real, or due to the mounting window, e.g. diamond?

❐ INSTRUMENT/SOFTWARE EFFECTS

The above-mentioned approach checks that all the bands and overall shape of the spectrum are not affected by the samples and the method of preparation; however, extra bands and anomalies may occur from instrument or software artefacts.

- Which laser line is used as a source, are resonance or self-absorption likely to affect band strengths?
- Does the spectrum really have a flat background or has a software background correction removed fluorescence, and destroyed information?
- Is the spectrum as strong as it appears? Check the scale and check for expansion routines.
- Has a smooth function been applied which leads to loss of bands normally resolved?
- Modern data systems display and plot information on data manipulation. Has this been applied? The lack of printed information does not mean manipulation has not occurred.
- Are the broad bands in the Raman spectrum due to fluorescence or burning?
- Are these sharp bands in the Raman spectrum which could come from cosmic rays or neon room lights?

❐ THE SPECTRUM

Once all the information on the sample history is acquired, and all possible distortions and artefacts have been identified, or dismissed, interpretation of the band positions and strengths should begin.

- Look at the total spectrum as a picture, does it look as expected from the sample. Are the bands broad or sharp? Are they strong or weak? Is the background sloping or flat? If it appears correct continue with band position interpretation.
- Start at the high wavenumber end, in the $3600-3100\,\mathrm{cm}^{-1}$ region; are there any –OH or –NH bands? Refer to Tables 1.1–1.5 to determine the type, and for confirmation, look for related bands in other parts of the spectrum, e.g. amides have carbonyl bands as well as –NH bands. These bands can be weak in Raman spectra and are easily missed or not seen.
- In the $3200-2700\,\mathrm{cm}^{-1}$ region, are there unsaturation or aliphatic bands present? Unsaturation is usually above $3000\,\mathrm{cm}^{-1}$, aliphatics below. If aliphatic bands are present, are they largely methyl or longer $-\mathrm{CH_2}-$ groups? Again refer to the tables for confirmation by other bands.
- Are these bands in the cumulative bond (e.g. $-\mathrm{N}{=}\mathrm{C}{=}\mathrm{N}$) region $2700-2000\,\mathrm{cm}^{-1}$?

- Are these bands in the double bond (e.g. $-C=O$, $-C=C-$) region $1800-1600\,cm^{-1}$? In the Raman spectrum unsaturated double bond bands are generally stronger and sharper than carbonyl bands. Infrared active bands can also appear in this region.
- By these checks, it should be established if the spectrum contains aliphatic, unsaturated or aromatic groups. Multiple bond bands or carbonyl bands should also have been identified. Look at the rest of the spectrum for strong bands. Do they correspond to bands in the tables?
- The region below $1600\,cm^{-1}$ contains many bands largely due to the finger-print of the molecule. Structural information can be gained from this region, but bands are mainly due to the backbone of the molecule. Selected phenyl ring modes and groups such as the azo group can be identified. Other groups with bands in this region tend to be oxygenated organics, e.g. nitro, sulpho, or heavily halogenated hydrocarbons. Inorganics have sharp Raman bands in this region (see tables in Chapter 6).
- Besides information identifying groups, is there negative information from bands that are not present? If the $3200-2700\,cm^{-1}$ region contains only very weak or no bands, then this negative information could be due to unusual species such as the halogenated species mentioned, that the Raman bands of these groups are too weak or that the sample is inorganic.
- Having established the possible groups present in the spectrum, can they be combined into a molecule which can be expected from the known chemistry and/or from the knowledge of possible impurities?

Always, wherever possible, crosscheck the interpretation by visibly matching to a reference spectrum of the molecule or of a very similar structure. Never trust peak list or computer search printouts without visually matching the spectra.

Finally check again if the answer makes sense with the sample. Is a red powder really ethanol? If this general procedure is followed, then the maximum information will be obtained from the examination, and errors will be minimized.

2.10.2 Computer Aided Spectrum Interpretation

Raman spectra can be interpreted for identification of substances by pattern matching, either by computer or by the hard work of visually searching through hard copy reference collections. There are now many commercial software packages and libraries available for rapid database searching of infrared spectra but still relatively few for Raman spectra. If computer searching is used, 'answers' must be crosschecked by visually comparing the sample and reference spectra. Do not rely on lists of nearest hits. Spectra can also be interpreted from first principles, by determining the chemical structural groups present from significant band positions, as stated previously. However, this type of interpretation is very difficult with Raman spectra. A knowledge of the type of chemistry involved is often a useful aid. Interpretation of spectra often requires experience and an understanding of the relevant answer required, particularly when mixtures or impure samples are examined.

Computer aided spectrum interpretation can be generally divided into two types. The most common is library searching or pattern matching. The other is structural elucidation, which is sometimes part of a training package or library search package for infrared spectra but is rarely found for Raman spectra.

❏ LIBRARY SEARCH SYSTEMS

Most Raman instrument manufacturers now offer their own library search routines which contain a few pre-recorded reference spectra and can be expanded with the user's own recorded spectra. Several can be enhanced with electronic versions of hard copy libraries such as the Aldrich and Sadtler^{TM} collections. The major library publishing companies also offer their own versions of electronic libraries as standalone library search systems. There is growth in Internet searchable libraries but Raman spectra are still very sparse.

❏ STRUCTURAL DETERMINATION AIDS

There have been many attempts at the development of artificial intelligence or expert systems to emulate the human thought process for spectral interpretation. Most have been successful for a limited range of similar chemical compounds or structures, but none have yet approached the full range attempted by humans. Many manufacturers of various spectroscopic instruments are including software training packages in their range of offerings, which very graphically demonstrate the fundamental principles of interpretation. Several now include excellent 3D graphic representation of band origins with simultaneous twisting, bending and stretching of bonds. These usually work very well for a limited range or group of molecules, but cannot be added to by the spectroscopist. As stated before Raman spectra are rarely included in this type of package.

An opposite approach is to use packages for DFT calculations to predict the band positions of various groups in the Raman spectrum of a molecule. These packages now can predict spectra which are very close to the spectra of actual samples. However it must be remembered that these packages generally assume single molecules in the vapour phase. Recorded spectra of solids and liquids can have band shifts due to molecular interactions.

❏ SPECTRA FORMATS FOR TRANSFER AND EXCHANGE OF DATA

Vibrational spectrocopists very quickly realized the potential of computers to manipulate spectra for quantitative or qualitative work. However this initially required using software supplied by the instrument manufacturer or difficult and tedious manipulation of the spectra files for transfer to another computer. With the advent of PC workstations and the establishment of large commercial databases, the demand grew for a universal format for data transfer. In 1987 the Joint Committee on Atomic and Molecular Physical Properties (JCAMP) proposed a format to be used internationally. This is known as JCAMP-DX. The format was intended to represent all data in a series of labelled ASCII fields of variable length. Very quickly the major instrument manufacturers provided software to convert their spectra to/from JCAMP format. Unfortunately whilst the data format was clearly specified, the file header format was less tightly specified. As a result commas, spaces, etc., were used in different ways as delimiters. The effect was that each manufacturer supplied a slightly different JCAMP file. A number of commercial spectrum file converters are now available which allow for the import and export of files from most spectrometers into data handling packages. The proliferation of Windows^{TM} based software has also removed the need for file transfer as the image of a spectrum can easily be transferred into reports and presentations using 'cut and paste' techniques.

❐ THE INTERNET

We are beginning to see what could become an explosion in the use of the Internet for spectroscopic information and assistance. All of the main instrument manufacturers have established home pages on the Internet. These are mainly used for promotional material but many have plans to provide access to notes on applications and give details of training courses or seminars. Most hard copy journals on spectroscopy are now available on the Internet for a fee. The *Society for Applied Spectroscopy* (*SAS*) journal has free monthly article listings. Recent editions have articles available in abstract form; however, the page numbers are not listed. Advance notices of meetings and events are also available. The *Internet Journal of Vibrational Spectroscopy* is free. It is a source of 'how to do it' articles, as well as the more formal journal articles. It contains a long list of links to relevant Internet sources. The free internet journal *Spectroscopy Now* has a specific Raman page.

2.11 SUMMARY

The advantages and disadvantages of Raman spectroscopy from a practical viewpoint are very clear from what is said in this chapter. It is extremely flexible and can be configured in many different ways. The continued improvements in modern optics including small diode lasers, improved simple detectors and fibre optic coupling have all led to the ability to use Raman scattering for problems for which we would not previously have considered it. Since it is a non-contact technique, it is possible to use it in a chemical factory with dust or inside the head of a combustion engine. Although the technique is limited by the fact that it is a weak effect, to some extent this can be overcome where the power density is high by the use of a microscope or particular forms of fibre optics. Thus, the future of Raman spectroscopy would appear to be set to advance particularly for specific analysis purposes. The disadvantage this creates is that the range of choice requires an understanding of the subject and cannot be made simply on the basis of the purchase of one simple instrument. However, most laboratories find that modern Raman instrumentation – visible or near-infrared FT systems – can solve many of the standard problems for which Raman scattering is deemed to be a suitable technique.

REFERENCES

1. P. Hendra, C. Jones and G. Warnes, *FT Raman Spectroscopy*, Ellis Horwood Ltd, Chichester, 1991.
2. B.T. Bowie, D.B. Chase and P. Griffiths, *Appl. Spectrosc.*, **54**, 200–207A (2000).
3. M.V. Pellow-Jarman, P.J. Hendra and R.J. Lehnert, *Vib. Spectrosc.*, **12**, 257–261 (1996).
4. H. Wang, C.K. Mann and J.V. Vickers, *Appl. Spectrosc.*, **56**, 1538–1544 (2002).
5. C.H. Chio, S.K. Sharma, P.G. Lucey and D.W. Muenow, *Appl. Spectrosc.*, **57**, 774–783 (2003).
6. B. Schrader and G.Z. Bergmann, *Anal. Chem.*, **225**, 230–247 (1967).
7. P.J. Hendra, *IJVS*, **1**, edition 1, section 1 (www.ijvs.com).
8. G. Dent, *Spectrochim. Acta A*, **51**, 1975 (1995).
9. Y.D. West, *IJVS*, **1**, edition 1, section 1 (www.ijvs.com).
10. K.J. Asselin and B. Chase, *Appl. Spectrosc.*, **48**, 699 (1994).
11. G. Dent and F. Farrell, *Spectrochim. Acta A*, **53**, 21–23 (1997).
12. C. Petty, *Vib. Spectrosc.*, **2**, 263 (1991).
13. N. Everall, *J. Raman Spectrosc.*, **25**, 813–819 (1994).
14. N. Everall and J. Lumsdon, *Vib. Spectrosc.*, **2**, 257–261 (1991).
15. J.S. Church, A.S. Davie, D.W. James, W.-H. Leong and D.J. Tucker, *Appl. Spectrosc.*, **48**(7), 813–817 (1994).
16. D. Louden, in: *Laboratory Methods in Vibrational Spectroscopy*, H.A. Willis, J.H. van der Mass and R.J. Miller (eds), John Wiley & Sons, Inc., New York, 1987.
17. M. Fleischmann, P.J. Hendra and A.J. McQuillan, *Chem. Phys. Lett.*, **26**, 163 (1974).
18. J.R. Lewis and P.R. Griffiths, *Appl. Spectrosc.*, **50**, 12A (1996).
19. S.M. Angel, T.F. Cooney and H. Trey Skinner, in: *Modern Techniques in Raman Spectroscopy*, J.J. Laserna (ed.), Ch. 10, John Wiley & Sons, Inc., New York, 2000.
20. D.A. Smith, S. Webster, M. Ayad, S.D. Evans, D. Fogherty and D. Batchelder, *Ultramicroscopy*, **61**, 247–252 (1995).
21. L. Song, S. Liu, V. Zhelyaskov and M.A. El-Sayed, *Appl. Spectrosc.*, **52**, 1364 (1998).
22. S.D. Schwab and R.L. McCreery, *Appl. Spectrosc.*, **41**, 126 (1987).

23. W. Xu, S. Xu, Z. Lu, L. Chen, B. Zhao and Y. Ozaki, *Appl. Spectrosc.*, **58**, 414–419 (2004).
24. N.J. Everall, *Appl. Spectrosc.*, **54**, 1515–1520 (2000).
25. N.J. Everall, *Appl. Spectrosc.*, **54**, 773–782 (2000).
26. B.R. Wood, S.J. Langford, B.M. Cooke, F.K. Glenister, J. Lim and D. McNaughton, *FEBS Lett.*, **554**, 247–252 (2003).
27. R.L. McCreery, *Raman Spectroscopy for Chemical Analysis*, Ch. 10, John Wiley & Sons, Inc., New York, 2000.
28. D.A. Carter, W.R. Thompson, C.E. Taylor and J.E. Pemberton, *Appl. Spectrosc.*, **49**, 11 (1995).
29. A.W. Fountain III, C.K. Mann and T.J. Vickers, *Appl. Spectrosc.*, **49**, 1048–1053 (1995).
30. NIST, www.cstl.nist.goc/div837/Division/techac/2000/RamanStandards.htm.
31. Kayser, www.kosi.com/raman/product/accessories/hca.html.
32. K.G. Ray and R.L. McCreery, *Appl. Spectrosc.*, **51**(1), 108–116 (1997).
33. R.L McCreery, www.chemistry.ohio-state.edu/~rmccreer/intensity/intesity.html.
34. B.T. Bowie, D.B. Chase and P. Griffiths, *Appl. Spectrosc.*, **54**, 164–173A (2000).

BIBLIOGRAPHY

This chapter contains a number of instructions and comments in sample preparation and use of accessories. These are not prescriptive, but based on experience, including getting it wrong. Sample specific and local needs will determine optimal conditions and procedures. To reference each original or milestone publication, invention, or application would have been too daunting a task; many have been gradually developed and improved by numerous workers over many years. Occasional references are given in the text to point the reader to greater in-depth understanding of the principles behind some of the necessarily brief descriptions. In addition to these, a short recommended bibliography relevant to this chapter is given.

J. Chalmers and P. Griffiths (eds), *Handbook of Vibrational Spectroscopy*, Vols 1 and 2, John Wiley & Sons, Inc., New York, 2001.
D.J. Gardiner and P.R. Graves (eds), *Practical Raman Spectroscopy*, Springer-Verlag, Berlin, Heidelberg, 1989.
J.G. Grasselli and B.J. Bulkin (eds), *Analytical Raman Spectroscopy*, John Wiley & Sons, Inc., New York, 1991.
ASTM, *1995 Annual Book of ASTM STDs, Vol.* 3.06, ASTM Philadelphia. (Designation E1683–95 Standard Practice For Testing the Performance of Scanning Raman Spectrometers.)

SOFTWARE INTERPRETATION TOOLS, DATABASES AND INTERNET SITES

Charles B. Adams, Colombia University, *IR Tutor*.
Biorad Laboratories, Sadtler Division, *IR Mentor, HaveIT all + new database*.

Chemical Concepts *Specinfo* Spectral Databases and *SpecTool* + *new database.*
Internet Journal of Vibrational Spectroscopy (http://www.ijvs.com).
Society for Applied Spectroscopy (http://www.s-a-s.org/journal/journal.htm).
Spectrochimica Acta (http://www.chemweb.com/gateways/elsevier.html).
Spectroscopy Europe (http://www.spectroscopyeurope.com).
Vibrational Spectroscopy (http://www.chemweb.com/gateways/elsevier.html).
Chemscape CHIME (http://www.mdli.co.uk/downloads/downloadable/index.jsp).
Raman Shift Frequency Standards (ASTME 1848) (http://chemistry.ohio-state.edu/
    ~rmccreer/shift.html#shiftdir).

Chapter 3

The Theory of Raman Spectroscopy

3.1 INTRODUCTION

As shown in Chapter 1, the sharp pattern of bands which make up a Raman spectrum makes it possible to use the technique for many types of analysis without a deep understanding of the nature of the effect. For example, it is possible to identify a molecule *in situ* from the pattern of bands and it may even be possible to determine the amount of the compound which is present. However a better understanding of the theory has real advantages. Much more information about a molecule and its surroundings can be obtained, the interpretation will be more secure, more possible pitfalls will be recognized and avoided, and the background required to understand some of the more exciting modern developments will be understood. There are many more detailed books on the theory of Raman spectroscopy but this chapter sets out and explains the salient points required for a more in-depth understanding. For example, where a mathematical treatment is required to make a specific point, the key equations are explained without a full derivation. The reader is referred to [1, 2] for a more thorough coverage.

Historically, Raman scattering has been described both in terms of 'classical theory' and 'quantum theory'. The older classical theory is based on the wave theory of light and is deficient in that it does not take into account the quantized nature of vibrations. In addition it is not able to explain as much about the relationship between molecular properties and Raman scattering as quantum theory. Thus, although this theory has persisted as an approach in many books, it is not described further here. Accounts of the theory can be obtained in [1, 2].

Modern Raman Spectroscopy – A Practical Approach   W.E. Smith and G. Dent
© 2005 John Wiley & Sons, Ltd   ISBNs: 0-471-49668-5 (HB); 0-471-49794-0 (PB)

3.2 ABSORPTION AND SCATTERING

When light interacts with matter, it can be absorbed or scattered. The process of absorption, discussed briefly in Chapter 1, requires that the energy of the incident photon corresponds to the energy gap between the ground state of a molecule and the excited state. It is the basic process used in a wide range of spectroscopic techniques and will be familiar to many readers. In contrast, scattering can occur whether or not there is a suitable pair of energy levels to absorb the radiation, and the interaction between the light and the molecule which causes this requires a different approach.

When a light wave, considered as a propagating oscillating dipole, passes over a molecule, it can interact and distort the cloud of electrons round the nuclei. This energy is released in the form of scattered radiation. Consider first the relative sizes of a light wave and a molecule. In the visible region, the wavelength of the light is between 400 and 700 nm whereas the size of a small molecule such as carbon tetrachloride is about 0.3–0.4 nm. Thus the oscillating dipole is much larger than the molecule. If it interacts with the molecule as it passes, it causes the electrons to polarize and go to a higher energy state. At that instant, the energy present in the light wave is transferred into the molecule. This interaction can be considered as the formation of a very short-lived 'complex' between the light energy and the electrons in the molecule in which the nuclei do not have time to move appreciably. This results in a high energy form of the molecule with a different electron geometry but without any large nuclear movement. This 'complex' between the light and the molecule is not stable and the light is released immediately as scattered radiation. It is often called the virtual state of the molecule. Since it has a different electronic geometry from that found in the static molecule and the nuclei do not have time to respond and reach a new equilibrium geometry to fit the distorted electronic arrangement, none of the electronic states of the molecule will describe the electron arrangement. Further the actual shape of the distorted electron arrangement will depend on how much energy is transferred to the molecule and hence is dependent on the frequency of the laser used. Thus, the laser defines the energy of the virtual state and the extent of the distortion. This virtual state is a real state of the transitory 'complex' formed.

The process differs from an absorption process in a number of ways. Firstly, the additional energy does not promote an electron to any one excited state of the static molecule; all states of the static molecule are involved to different extents and are mixed together to form states of the distorted 'complex'. The energy of this state is dependent on the energy of the laser used and the amount of distortion is dependent on the electronic properties of the molecule and on the energy of the laser. Secondly, the lifetime of the excited state is very short compared to most absorption processes. The radiation is scattered as a sphere and not lost by energy transfer within the molecule or emitted at a lower energy. Thirdly, and this will be dealt with later in this chapter, there is a link between the polarization of the exciting and scattered photons which can be of value in assigning particular vibrations.

Two types of scattering are readily identified. The most intense form of scattering, Rayleigh scattering, occurs when the electron cloud relaxes without any nuclear movement. This is essentially an elastic process and there is no appreciable change in energy. Raman scattering on the other hand is a much rarer event which involves only one in 10^6–10^8 of the photons scattered. This occurs when the light and the electrons interact and the nuclei begin to move at the same time. Since the nuclei are much heavier than the electrons, there is an appreciable change in energy of the molecule to either lower or higher energy depending on whether the process starts with a molecule in the ground state (Stokes scattering) or from a molecule in a vibrationally excited state (anti-Stokes scattering). Figure 1.2 in Chapter 1 shows a simple diagram illustrating Rayleigh and Raman scattering. In each case the energy of the virtual state is defined by the energy of the incoming laser. The two states marked m and n are different vibrational states of the ground electronic state.

Most molecules at rest prior to interaction with the laser and at room temperature are likely to be in the ground vibrational state. Therefore the majority of Raman scattering will be Stokes Raman scattering. The ratio of the intensities of the Stokes and anti-Stokes scattering is dependent on the number of molecules in the ground and excited vibrational levels. This can be calculated from the Bolzmann equation,

$$\frac{N_n}{N_m} = \frac{g_n}{g_m} \exp\left[\frac{-(E_n - E_m)}{kT}\right] \tag{3.1}$$

N_n is the number of molecules in the excited vibrational energy level (n),
N_m is the number of molecules in the ground vibrational energy level (m),
g is the degeneracy of the levels n and m,
$E_n - E_m$ is the difference in energy between the vibrational energy levels,
k is Boltzmann's constant ($1.3807 \times 10^{-23} \, \text{JK}^{-1}$).

We shall see, when we consider symmetry later in this chapter, that some vibrations can occur in more than one way and the energies of the different ways are the same, so that the individual components cannot be separately identified. The number of these components is called the degeneracy and is given the symbol g in Equation (3.1). Since the Boltzmann distribution has to take into account all possible vibrational states, we have to correct for this. For most states g will equal 1 but for degenerate vibrations it can equal 2 or 3.

3.3   STATES OF A SYSTEM AND HOOKE'S LAW

Any molecule consists of a series of electronic states each of which contains a large number of vibrational and rotational states. In Figure 3.1 a sketch of a typical ground electronic state of a molecule is shown. The y-axis represents the energy of the system and the x-axis the internuclear separation. The curved line represents the electronic state. At large internuclear separations, the atoms are essentially free and as the distance decreases they are attracted to each other to form a bond. If they approach too closely, the nuclear forces cause repulsion and the energy of the molecule rises steeply as shown. Thus the lowest energy is at the bond length. However within the curve, not every energy is possible since the molecules will be vibrating and the vibrational energies, which are quantized, have to be taken into account. The tie lines are the quantized vibrational states. A particular vibrational level of a particular electronic state is often called a vibronic level.

At first glance this curve, referred to as a Morse curve, is relatively simple but there are more complications which are generally not added because the diagram gets too cluttered for use. What is shown in the figure refers to one vibration. The first level ($v = 0$) is the ground state where the molecule is not vibrating and the second level ($v = 1$) is the state where one quantum of the correct energy is absorbed and the molecule vibrates. The levels above this require energies of approximately but not exactly two times, three times, four times, etc., of the quanta required to move the molecule from the ground state 0 to the first excited state 1. Where a change of more than one quantum occurs the peak obtained is called an overtone. As we shall see, in Raman scattering this occurs only in special circumstances. In most Raman spectra overtones are predicted as very weak or non-existent. To describe all the vibrations in a molecule such as in Figure 3.1, a similar set of tie lines but at different energies

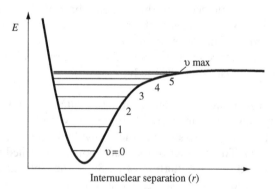

Figure 3.1.   A typical Morse curve for an electronic state showing the vibrational levels as horizontal tie lines.

is required for each vibration. Further, vibrations can combine so that one quantum of one vibration and one of another vibration will give a new level. In the spectrum, peaks due to these combinations are called combination bands and like overtones appear only in certain circumstances. To make matters even more complicated, rotational levels, which are of lower energy than vibrational levels, also require to be added. A diagram with all these levels is too complex to use and conventionally is simplified either by showing all the levels for one vibration or one vibrational level for each vibration depending on the use to which the diagram will be put.

To describe the process of absorption when an electron is excited from one electronic state to another, a Morse curve for the ground and excited state is required with the excited state plotted above the ground state since it will be at higher energy. In Raman scattering, as we shall see later in the chapter, all excited vibronic states have an influence on scattering efficiency. As a result, in principle, we require to draw Morse curves for all states of the molecule. However, the influence of each state is not specific and hence a simpler diagram as shown in Figure 1.2 can be used in which all the many excited vibronic levels from the many excited electronic states are represented by a few lines. Further, since Raman scattering is fast compared to the time for nuclear movement, there is no appreciable change in the nuclear separation during any one scattering event and therefore no change along the x-axis. Thus, for a simple description of the process, energy changes in the molecule are plotted as vertical lines and states as horizontal lines with the other features of the Morse curve neglected.

By way of revision, it is useful to remind ourselves of the main features of Figure 1.2. It shows the energy changes which occur when the exciting radiation interacts with the molecule to form a 'virtual' state and the scattering which follows when the molecule relaxes. The scattered radiation is what we measure as Raman scattering and the energy difference between the excitation and scattering processes corresponds to the energy of vibrations of the molecule. As we shall see below, the two levels can vary by only one quantum number for fundamentals. There are features in Figure 1.2 which are potentially misleading. The y-axis is an energy axis. However, in Raman scattering the energy of a C—C vibration may typically be between 1600 and 1000 cm^{-1} but the energy of a green laser will be about 20,000 cm^{-1}. Very often, as here, the energy of the laser radiation is not given accurately because the desire is to show the vibration spacing clearly, and plotting the true excitation energy would lead to a very large separation between the ground state and the virtual state reducing the space to show the vibronic levels.

The shape of the Morse curve makes it difficult, but not impossible, to calculate the energy of vibronic levels and so simple theory uses the harmonic approximation. In this approach, the Morse curve shown is replaced by a parabola calculated for a diatomic molecule by considering it as two masses connected by a vibrating spring.

With this approach, Hooke's law (Equation (3.2)) gives the relationship between frequency, the mass of the atoms involved in the vibration and the bond strength for a diatomic molecule:

$$\nu = \frac{1}{2\pi c}\sqrt{\frac{K}{\mu}} \qquad (3.2)$$

where c is the velocity of light, K is the force constant of the bond between A and B, and μ is the reduced mass of atoms A and B of masses M_A and M_B;

$$\mu = \frac{M_A M_B}{M_A + M_B} \qquad (3.3)$$

Hooke's law makes it easy to understand the approximate order of the energies of specific vibrations. The lighter the atoms, the higher the frequency will be. Thus C–H vibrations lie just below and just above $3000\,cm^{-1}$ and C–I vibrations at less than $500\,cm^{-1}$. The force constant is a measure of bond strength. The stronger the bond, the higher the frequency will be. A list of vibrational energies is given in Chapter 1.

Two other points should be noted. The harmonic approximation predicts that the overtones of a molecule are equally spaced but the reality is that the departure from harmonicity in a real system will mean that, particularly at higher energies, the energy separations between levels will decrease. For example, in Figure 3.1 which shows a Morse curve for one vibration, all the vibrational levels are shown as they actually are, with a decreased separation in energy the higher the vibrational quantum number. They would be shown equally spaced in the harmonic approximation. Further, the electron density along the vibration is of importance in working out the efficiency of the Raman process. We will make use of this later in discussing resonance in Chapter 4.

3.4 THE NATURE OF POLARIZABILITY AND THE MEASUREMENT OF POLARIZATION

When radiation is emitted from a source, a number of photons are emitted and each photon consists of an oscillating dipole. Observed at 90° to the direction of propagation, the beam looks like a wave. Observed looking along the line between the observer and the light source, each photon will appear as a line, with the oscillating dipole in that line. In general the angle of the line to the observer is random, but by passing the light through a suitable optical element such as a Nichol prism or a piece of Polaroid film, all the lines can be made to propagate in one direction. This is called plane or linearly polarized radiation. The lasers that are normally used for excitation in Raman scattering are usually at least partially polarized. Good Raman spectrometers also have an optical element, a polarizer, that can be put in the beam to ensure that the light is linearly polarized.

When linearly polarized light interacts with the molecule, the electron cloud is distorted by an amount that depends on the ability of the electrons to polarize (i.e. the polarizability, α). The light causing the effect is polarized in one plane, but the effect on the electron cloud is in all directions. This can be described as a dipole change in the molecule in each of the three Cartesian co-ordinates x, y and z. Thus, to describe the effect on molecular polarizability of an interaction with linearly polarized radiation, three dipoles require to be considered. The simple expression is that a dipole μ is created in the molecule by the field from the incident photon E.

$$\mu = \alpha E \tag{3.4}$$

To allow for the polarization angle of the linearly polarized light, the polarizability components of the molecule are usually labelled, an example of which is shown below:

$$\alpha_{xx}$$

The first subscript x refers to the direction of polarisability of the molecule, and the second x refers to the polarization of the incident light. Thus, $\mu_x = \alpha_{xx}E_x + \alpha_{xy}E_y + \alpha_{xz}E_z$. Similar expressions will exist for both μ_y and μ_z.

Thus, the polarizability of the molecule is a tensor,

$$\begin{bmatrix} \mu_x \\ \mu_y \\ \mu_z \end{bmatrix} = \begin{bmatrix} \alpha_{xx} & \alpha_{xy} & \alpha_{xz} \\ \alpha_{yx} & \alpha_{yy} & \alpha_{yz} \\ \alpha_{zx} & \alpha_{zy} & \alpha_{zz} \end{bmatrix} \begin{bmatrix} E_x \\ E_y \\ E_z \end{bmatrix}$$

There are specific advantages of this rather complex arrangement. In Raman scattering the incident and scattered beams are related. If radiation of a particular polarization is used to create the Raman scattering, the polarization of the scattered beam is related to but not necessarily the same as that of the incident beam. Thus, a Raman spectrometer has an optical element, the polarizer, to control the polarization of the incident beam. It ensures that the radiation is plane polarized and determines the angle of the plane of the incident radiation. A second element, the analyser, analyses the polarization of the scattered beam. The analyser works by allowing the polarized light to pass through to the detector only in one plane. It is initially set to allow transmission of scattered radiation in the plane of the incident radiation (called

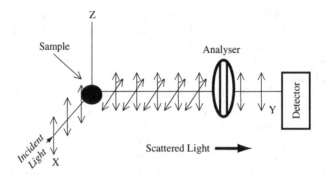

Figure 3.2. Arrangement to monitor polarization of Raman scattering. The arrows indicate the plane of the scattered light. The analysis is set to allow through only parallel scattering. If rotated 90° it will allow through only perpendicular scattering.

parallel scattering). It is then set at 90° to this direction to allow any light in which the polarization direction has been changed by the molecule to pass through to reach the detector (called perpendicular scattering). This arrangement is shown in Figure 3.2.

If a single crystal is used as a sample, all molecular axes are lined up within the unit cell in the same direction for each cell. Thus, the polarization direction of the incident radiation bears a relationship to the molecular axes. With an arrangement like this, it is possible to analyse each of the components of the tensor shown above. This works best for crystals in higher symmetry space groups but not cubic. Light is a dipole property, which means that the optical axes of a material are set at 90° to each other. In some higher symmetry space groups such as the tetragonal space group, the optical and crystal axes are at right angles and so they can be aligned to match the polarization direction of the incident beam. Under these circumstances, light polarized in the z direction, passing through the crystal along the z-axis, will pick out the component α_{zz}. In all likelihood there will be a molecular axis along the z-axis and so the information can be related to molecular properties. However in most situations the analysis is more complex. Light which is not sent down an axis of the crystal will rotate within it and in many crystal space groups, the crystal axes are not at right angles and bear a complex relationship to the molecular axes. Thus this approach is very informative for a very limited number of samples. It is not dealt with further here but the reader should be aware of the possibility that, when using single crystal samples, the intensities of the bands may be affected by the structure.

Often, the samples we examine are either in the gas phase or in solution. In either case there is no ordering of the axes of the molecule to the polarization direction of the light but information can still be obtained from polarization measurements. What is measured in practice is the depolarization ratio where the intensity of a given peak is measured with the plane of polarization of the incident light parallel or perpendicular to the scattered light analysed. For samples such as this, it is useful to express the average polarizability in terms of two separate quantities that are invariant to rotation, namely isotropic and anisotropic scattering. Isotropic scattering is measured with the analyser parallel to the plane of the incident radiation and anisotropic scattering with the analyser perpendicular to the plane. It is possible to solve the tensor and calculate the ratio of parallel to perpendicular scattering (see [1]). This ratio is what is actually measured. It is called the depolarization ratio (ρ). Here we illustrate the salient equations but do not give details since this ratio is usually used qualitatively and is often talked about but seldom calculated.

The isotropic and anisotropic parts of the tensor are represented in Equations (3.5) and (3.6),

$$\bar{\alpha} = \frac{1}{3}(\alpha_{xx} + \alpha_{yy} + \alpha_{zz}) \tag{3.5}$$

$$\gamma^2 = \frac{1}{2}\left[(\alpha_{xx} - \alpha_{yy})^2 + (\alpha_{yy} - \alpha_{zz})^2 + (\alpha_{zz} - \alpha_{xx})^2 + 6\left(\alpha_{xy}^2 + \alpha_{xz}^2 + \alpha_{yz}^2\right)\right] \tag{3.6}$$

and the effect on parallel and perpendicular polarization of Equations (3.7) and (3.8),

$$\bar{\alpha}_{\parallel}^2 = \frac{1}{45}(45\bar{\alpha}^2 + 4\gamma^2) \tag{3.7}$$

$$\bar{\alpha}_{\perp}^2 = \frac{1}{15}\gamma^2 \tag{3.8}$$

This gives a ratio between parallel and perpendicular scattering as

$$\rho = \frac{\bar{\alpha}_{\perp}^2}{\bar{\alpha}_{\parallel}^2} = \frac{3\gamma^2}{45\bar{\alpha}^2 + 4\gamma^2} \tag{3.9}$$

The importance of this information becomes clear only when we consider the selection rules for Raman scattering later in the chapter. In essence, for a molecule with appreciable symmetry in solution or in the gas phase, the depolarization ratio varies depending on the symmetry of the vibration. Symmetric vibrations have the lowest depolarization ratios. Thus measurement of parallel and perpendicular scattering using the analyser to obtain the depolarization ratio provides a check on assignments of the peaks. This check is not available with absorption spectroscopies such as infrared.

There is one final practical point which has to be borne in mind. When radiation from the analyser is detected via a monochromator, the efficiency of the grating used to split up the light is dependent on the plane of polarization. This means the grating will transmit radiation to the detector more efficiently for either parallel or perpendicular polarization and consequently the apparent depolarization ratio will be wrong. The most conventional way to overcome this problem is to add an extra element, a scrambler, which scrambles the polarization of the light before it enters the monochromator so that the detector is equally efficient for all polarization directions of the incoming radiation. There are other ways of doing this. For example a half-wave plate can be inserted instead. This rotates the light by 90° and is swung into the beam in only one direction of the analyser so that the light in both analyser positions enters the monochromator in the same direction.

Failure to appreciate this effect can be serious. Laser radiation is usually linearly polarized to a significant extent. In the absence of polarizers and analysers and if no scrambler is in place, the laser acts as the polarizer and the monochromator as the analyser. This means that spectra often labelled as 'unpolarized' because no polarization optics were used will be polarized. As a result, the intensities can be misleading, particularly for molecules with high symmetry.

3.5 THE BASIC SELECTION RULE

The basic selection rule is that Raman scattering arises from a change in polarizability in the molecule. As we shall demonstrate later, this means that symmetric vibrations will give the most intense Raman scattering. This is in complete contrast to infrared absorption where a dipole change in the molecule gives intensity and, at a very simple level, this means asymmetric rather than symmetric vibrations will be intense.

3.6 NUMBER AND SYMMETRY OF VIBRATIONS

With any molecule, the energy can be divided into translational energy, vibrational energy and rotational energy. Translational energy can be described in terms of three vectors 90° to each other and so has three degrees of freedom. Rotational energy for most molecules can also be described in terms of three degrees of freedom. However, for a linear molecule there are only two rotations. The molecule can either rotate around the axis or about it. Thus, molecules are said to have three translational degrees of freedom and three rotational degrees of freedom with the exception of linear molecules, which have two degrees of rotational freedom. All other degrees of freedom will be vibrational degrees of freedom and each is equivalent to one vibration. There-fore, the number of vibrations to be expected from a molecule with N atoms is $3N - 6$ for all molecules except linear systems where it is $3N - 5$.

From this it is possible to work out the number of vibrations which occur. However, it must be noted that this does not make the vibrations either Raman or infrared active and in general we would not expect to observe all vibrations in either spectroscopy.

As discussed in Chapter 1, for a simple diatomic molecule, which by defini-tion is linear, there is one vibration. For a simple homonuclear diatomic like oxygen or nitrogen this is a symmetric vibration in which we would not expect any infrared activity, but we would, since the bond is stretched, expect a change in polarizability to occur. Thus, one band would be expected in the Raman spectrum and there would be no band in the infrared spectrum.

When a molecule has a number of symmetry elements in its structure, more selection rules apply. Consider a square planar molecule such as $AuCl_4^-$ which is illustrated with selected vibrational movements indicated by arrows in Figure 3.3. This molecule is said to have a centre of symmetry. The definition of a centre of symmetry is that any point in the molecule reflected through the central point will arrive at an identical point on the other side. Thus, in this molecule, ignoring vibrational movement, any chlorine atom reflected through the gold centre will arrive at an identical chlorine atom on the other side. An example of a molecule that does not possess this property is the nitrate ion in which an oxygen reflected through the nitrogen in the centre would arrive at a point in space (see Figure 3.4).

We need to have some way of describing the vibrations in a molecule. In principle, it would be possible to use x, y and z co-ordinates and simply explain how each atom moves by how much these co-ordinates change. This would be complicated and we would not understand the nature of the information read-ily. The usual way is to use normal co-ordinates as shown for $AuCl_4^-$ in Figure 3.3. Normal co-ordinates of a molecule make use of the natural directions of bonds and are those co-ordinates in which all atoms vibrating go through the centre of gravity of the molecule at the same time. The value of normal co-ordinates is that they provide a much better visual pattern of what a vibra-tion looks like. Two vibrations for the $AuCl_4^-$ molecule using normal co-ordinates are illustrated in Figure 3.3.

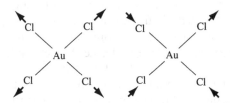

Figure 3.3. Illustration of two vibrations in the centrosymmetric ion $AuCl_4^-$.

In one vibration, all the atoms move out at the same time, and in the other, two atoms move in as two move out. The arrows representing the movement of atoms could also be reflected through the centre in the same way as atoms. The assignment of the molecule as centrosymmetric was based on the properties of the atoms with the molecule at rest in its equilibrium position. The vibrational movements do not affect that. However, clearly there is a difference between the two vibrations as shown in Figure 3.3. For this reason, an extra label is used. Vibrations of the first type are called even, or gerade, and are labelled g, whereas those of the second type are odd, or ungerade, and are labelled u. This labelling applies only to molecules with a centre of symmetry. It will not, for example, apply in nitrate.

For any molecule with symmetry elements, it is possible to use symmetry to help understand molecular motion by applying group theory. The approach is very powerful in experiments involving molecules of high symmetry, and gives a better insight into selection rules. In addition, the use of labels which arise from group theory is common throughout the literature. To enable the reader to understand its value, the basics of the approach are described below. However, it is often not useful in experiments with more complex molecules and so an extensive treatment is outside of the scope of this book. Good texts such as the book by Cotton [3] describe the application of symmetry in detail.

3.7   SYMMETRY ELEMENTS AND POINT GROUPS

Any molecule can be classified by its symmetry elements (i.e. axes and planes). It is then possible to assign the molecule to a group called a point group which has these same elements. This can then be used to predict which bands are infrared and which are Raman active. To do this it is necessary to work out the symmetry elements in the molecule. The main symmetry elements we need to recognize are the following:

E – The identity element. This takes the molecule back into the same position it started from; i.e., a 360° rotation for every part of the molecule does this.

C_n – An axis of symmetry in which the molecule is rotated about a molecular axis. n Defines how often the molecule requires to be rotated to arrive back at the starting point. Thus, in the nitrate ion shown in Figure 3.4, one possible axis is the one pointing straight out of the plane of the paper. If the molecule is rotated about it, each oxygen will require to be rotated three times to arrive back at the starting point. This is a C_3 axis. There may well be a number of axes in a molecule. For example in the nitrate ion, three C_2 axes also exist. They lie along the NO bonds and rotating the molecule about them would require two rotations to take the molecule back to its starting point. The axis with the highest value of n, for the nitrate C_3, is known as the principle axis of the molecule.

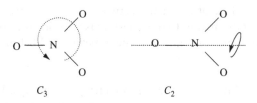

Figure 3.4. C_3 and C_2 axes in the nitrate ion.

σ_h – A plane of symmetry in which the plane is perpendicular to the principle axis of the molecule.

σ_v – A plane of symmetry in which the plane is parallel to the principle axis of the molecule.

i – A centre of inversion in which every point inverted through the centre arrives at an identical point on the other side.

S_n – An axis which combines a rotation and an inversion.

These symmetry elements define a particular type of molecule. All molecules with the same set of symmetry elements are said to belong to the same point group.

To assign a molecule to its point group, the symmetry elements are first recognized and then analysed according to a set of rules. Usually symmetry elements are not analysed to assign a molecule to a particularly high symmetry point group such as the cubic point group, the octahedral point group and the tetrahedral point group. These can usually be recognized immediately. The questions we ask to make an assignment are set out in order below:

1. What is the principle axis of the molecule?
2. Is there a set of n C_2 axes at right angles to it? If the answer is no, carry on with the questions below. If the answer is yes, go to question 6.
3. Is there a plane perpendicular to the principle axis? If so, this is a σ_h plane. A molecule which has a C_n principle axis and a σ_h plane can be assigned to the point group C_{nh}.
4. If there is no σ_h plane, are there planes of symmetry parallel to the principle axis? There should be as many planes of symmetry as the n value. If this is the case, the point group is assigned as C_{nv}.
5. If there are no planes, the point group is assigned as C_n.
6. If the molecule has a principle axis and a set of n C_2 axes at right angles to it, is there a plane of symmetry perpendicular to the principle axis (i.e. a σ_h plane)? If this is the case, this molecule belongs to the D_{nh} point group.

7. If there is no σ_h plane of symmetry, is there a set of n σ_v planes parallel to the principle axis? If the answer to this question is yes, then the point group is D_{nd}.
8. If there are no planes of symmetry, the molecule will belong to the D_n point group.

Other molecules will belong to lower symmetry point groups. For example, for some molecules there is an S_n axis or a σ_v plane or no symmetry element at all. These can be recognized and a point group assigned by inspection.

Having assigned the molecule to a point group, group theory can be used to predict whether or not a band will be Raman or infrared active. It is particularly important to note that symmetry considerations allow us to determine whether or not a band is allowed in a Raman or infrared spectrum. This does not tell us how strong it will be; this would require a calculation.

There is a group theory table for all point groups which defines the symmetry behaviour of every vibration of a molecule belonging to that point group. Below, we reproduce the C_{2v} point group table which would be the correct point group for a single molecule of water.

C_{2v}	E	C_2	$\sigma_v(xz)$	$\sigma_v'(yz)$		
A_1	1	1	1	1	z	x^2, y^2, z^2
A_2	1	1	-1	-1	R_z	xy
B_1	1	-1	1	-1	x, R_y	xz
B_2	1	-1	-1	1	y, R_x	yz

In the table, the symmetry elements are shown across the top. The first column contains a series of letters and numbers. The first one we see is A_1. This is a way of describing a vibration, or for that matter an electronic function. It describes what happens to the vibration with each symmetry element of the molecule. These symbols are called irreducible representations and the top line always contains the one which refers to the most symmetrical vibration in terms of its behaviour when it is rotated or reflected by the symmetry operations. In higher symmetry point groups where there is a centre of symmetry, there would also be a g or a u subscript. For example, the most symmetric representation in the D_{4h} point group to which the molecule $AuCl_4^-$ belongs is A_{1g}. There are four possible letters, A, B, E and T. A and B mean that the vibration is singly degenerate. E means it is doubly degenerate and T means it is triply degenerate. In the C_{2v} point group all vibrations are singly degenerate. A is more symmetric than B. Across the line from the symbols representing the irreducible representations, there are a series of numbers for each. The numbers are either 1 or -1 and 1 is more symmetric than -1. For example, in the table, an A_1 irreducible representation gives the value of 1 for every symmetry element.

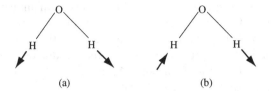

Figure 3.5. Two vibrations of water.

Figure 3.5 shows two vibrations of water. By looking at the shape of the molecule it is possible to assign it to the point group C_{2v} using the methods given above. For the vibration in (a), when the molecule is rotated about the C_2 axis, the direction of the arrow representing a vibration does not change. This is the highest symmetry and is denoted as 1. In addition the direction does not change when the arrow is reflected by either of the planes of symmetry (the plane of the paper and one perpendicular to it which bisects the oxygen). Therefore, the vibration (a) is assigned to the highest symmetry irreducible representation of the C_{2v} point group (A_1). In vibration (b) the sign of the arrow is reversed for C_2 and one plane. When this happens this is given the number -1. Thus, vibration (b) belongs to a lower symmetry representation. The actual label depends on which plane of symmetry is considered first in the table. It is conventionally given the irreducible representation B_1.

By this method we can assign a vibration to a particular irreducible representation in a particular point group. For more complex molecules there is a procedure to follow to do this and this is explained in books on the subject [3].

The main advantage of this assignment is that the tables also contain information that enables us to work out whether the vibrations will be allowed by symmetry or not. For infrared, this is done by multiplying the irreducible representation of the vibration by the irreducible representation of x, y or z which is given in the end column of the point group table in most, but not all, layouts. These correspond to three Cartesian co-ordinates of the molecule and are the irreducible representations of the dipole operator. If this result contains the totally symmetric representation (the highest symmetry representation in a particular point group A_1 in the point group C_{2v} but A_{1g} in the point group D_{4h}) then the vibration is allowed. The reason this works is that a vibration can be allowed only if the product of the irreproducible representations of the ground state, the operator, and the excited state is totally symmetric. Since the ground state is always totally symmetric, it turns out that we only need to multiply the other two. A similar approach is adopted for Raman scattering but in this case we look for the more complex quadratic functions x^2, y^2, z^2, xy, $x^2 - y^2$, etc., in the table and these are multiplied by the symmetry representation of the vibration. For simple point groups with non-degenerate representations, the rules for multiplying irreducible representations are $A \times A = A$, $B \times B = A$, $A \times B = B$, $1 \times 1 = 1$, $2 \times 2 = 1$ and $1 \times 2 = 2$.

It is unlikely that many readers will carry out an in-depth analysis of this type and in the interests of balance, the reader who requires more information is referred to a group theory book for a fuller explanation [3]. However, the symbols are often used in spectroscopy and the irreducible representations can be used to show if a band is allowed. All readers with serious interests in spectroscopy need to know what they mean.

3.8   THE MUTUAL EXCLUSION RULE

One crucial result which arises from this analysis is that irrespective of other symmetry considerations, for a centrosymmetric molecule, only vibrations which are g in character can be Raman active and only vibrations which are u in character can be infrared active. This is because irrespective of the exact irreducible representation, the g and u labels can be multiplied out and the final product must contain the totally symmetric representation and hence g. The rules are $g \times g = g, u \times u = g$ and $g \times u = u$. Since the Raman operators are g in character and the ground state is g, the excited state must be g if the vibration is to be allowed. In contrast, the infrared operator is u in character and so the excited state must be u if the vibration is to be allowed. Thus, in a molecule with a centre of symmetry, vibrations which are Raman active will not be infrared active, and vibrations which are infrared active will not be Raman active. Note that, as stated in Section 3.2 without proof, it is the symmetric vibrations (g) which are Raman active, and the asymmetric vibrations (u) which are infrared active. This analysis leads to a rule known as the mutual exclusion rule, which states that any vibration in a molecule containing a centre of symmetry can be either Raman or infrared active, but not both. In molecules without a centre of symmetry, there is no such specific rule. Nonetheless, in general, symmetric vibrations are more intense in Raman scattering and asymmetric vibrations in infrared scattering.

3.9   THE KRAMER HEISENBERG DIRAC EXPRESSION

The development of the theory of light scattering is beyond the scope of this book. What we will do here is choose two of the key equations from light scattering theory and develop them to understand in more depth the theory of Raman scattering [4–6]. The intensity of Raman scattering is defined by Equation (3.10):

$$I = Kl\alpha^2\omega^4 \tag{3.10}$$

K consists of constants such as the speed of light, l is the laser power, ω the frequency of the incident radiation and α the polarizability of the electrons in the molecule. Thus, two of the parameters which are variable are under the control of the spectroscopist, who can set the laser power and the frequency of the incident light. The way in which l and ω are used to maximize the potential of Raman scattering has already been considered (Chapter 2). The theory is required to understand the role of the molecular property, the polarizability α.

The equation used to describe polarizability in the molecule is known as the Kramer Heisenberg Dirac (KHD) expression. It is a large equation but it can be easily understood with little in the way of mathematical knowledge. All the terms are defined below and the process being described is the one shown diagrammatically in Figure 1.2:

$$\left(\alpha_{\rho\sigma}\right)_{GF} = k \sum_{I} \left(\frac{\langle F|r_{\rho}|I\rangle\langle I|r_{\sigma}|G\rangle}{\omega_{GI} - \omega_{L} - i\Gamma_{I}} + \frac{\langle I|r_{\rho}|G\rangle\langle F|r_{\sigma}|I\rangle}{\omega_{IF} + \omega_{L} - i\Gamma_{I}} \right) \qquad (3.11)$$

α is the molecular polarizability and ρ and σ are the incident and scattered polarization directions. Σ is the sum over all vibronic states of the molecule as might be expected from the non-specific nature of scattering. Outside this the remaining terms are constants. G is the ground vibronic state, I a vibronic state of an excited electronic state and F the final vibronic state of the ground state. G and F are simply the initial and final states of the Raman scattering process as shown in Figure 1.2. We will consider the numerator and the denominator separately and define the terms in the denominator in due course.

To understand the numerator, consider the numerator in the first term. It consists of two integrals. Because of the complexity of the expression it is usual to write the integrals using 'bra' and 'ket'($\langle|$ and $|\rangle$) nomenclature rather than standard integrals. These integrals are similar to those used in electronic adsorption spectra to describe the absorption and emission processes but since light is not promoted to any actual state of the molecule in Raman scattering, they are better considered as terms which mix the ground and excited states in order to describe the distorted electron configuration in the complex between the molecule and the light. One of the integrals is shown below:

$$\langle I|r_{\sigma}|G\rangle$$

Starting from the right-hand edge of the expression, $|G\rangle$ is a wave function to represent the ground vibronic state of the ground electronic state. The operator r_{σ} is the dipole operator and the mathematical process of it operating on $|G\rangle$ and multiplying the product with the excited state $\langle I|$ mixes the two states and when the result is summed over all states. This describes in part the excitation process. A similar process describing in part the scattering process occurs with the left-hand integral to leave the molecule in the final state $\langle F|$. Thus, the first of the two triple intervals mixes a ground and an excited state and the second of these integrals mixes the excited state and the final state. Since it is a mixing between two states which is being described, there is no reason why this process should start in the ground state. Thus, in the second term in Equation (3.11), an equivalent expression to that in the first term is added. This starts with the excited state and mixes the excited and ground states together in the same way. Fortunately, as we shall see when we come to the denominator, this term is less significant in Raman scattering.

Earlier, the nature of the virtual state was discussed and it was pointed out that a virtual state is a real state of the distorted molecule but, since the nuclei do not have time to reach equilibrium, it is not any state of the static molecule. Thus, when the KHD expression is used, the process of distortion is described by mixing all of the vibronic, excited and ground states together to describe electronic states of the molecule which exist only for the instant in which the light is captured. Consider the denominators of terms 1 and 2 in the expression in 3.11. The energy of the term $i\Gamma_I$ is small compared to the energies ω_{GI} and ω_L. In term 1, the nearer a specific excited state I is in energy, to the low energy the smaller the denominator is and the larger part the particular expression for that state will play in the final expression. Further, because ω_{GI} and ω_L are added in the second expression, the denominator will always be large compared to that in the first term. Consequently term 2 plays a smaller role in describing the polarization process and will now be neglected.

Without $i\Gamma_I$, when the frequency of the incident laser light is the same as the frequency of an electronic transition, then the denominator of the first term would go to zero and the result would be that the scattering would become infinite! The term $i\Gamma_I$ relates to the lifetime of the excited state and affects the natural breadth of Raman lines. Thus, although it is small it is a vitally important part of the basic equation defining molecular polarizability.

Each of the expressions in the numerator of term 1 will depend on the exact nature of the states and the way in which they are coupled through the operator. In this paragraph, we further analyse the KHD expression particularly to understand the selection rules in Raman scattering and to lay the foundation for the resonance Raman approach in Chapter 4. To do this, the states are usually split up into electronic and vibrational components using the Born Openheimer approach. In this approach, the total wave function is split up into separate electronic (θ), vibrational (Φ) and rotational (r) components.

$$\Psi = \theta \cdot \Phi \cdot r \qquad (3.12)$$

This is a very successful way of approaching many spectroscopy problems. It works because of the difference in the timescale of electronic, vibrational and rotational transitions. The very light electrons involved in a pure electronic transition will change from a ground to an excited state in a timescale in which there is very little movement of the nucleus (10^{-13} or less of a second). This is the reason electronic transitions are drawn vertically in the conventional diagrams such as Figure 1.2. The distance along the x-axis which plots internuclear separation cannot alter appreciably during the transition. Vibrational transitions occur in about 10^{-9} of a second and are faster than rotational transitions. Although rotational effects can be seen in gas phase Raman spectra, for the purposes of the limited theory given here, the rotational contribution will be largely neglected.

Thus, because of the different timescales, electronic and vibrational terms can be separated. The term θ is the electronic part of the expression and will depend on both the nuclear and electronic co-ordinates (R and r respectively) whereas the vibrational term which involves displacement of the heavier nuclei will depend entirely on the nucleic co-ordinates (R). The separation between the vibrational and electronic functions allows the integrals in the numerator in the KHD expression to be split up. The electrons travel from one excited electronic state to another so only the electronic term involves the operator.

$$\langle I|r_\sigma|G\rangle = \langle \theta_I \cdot \Phi_I|r_\sigma|\theta_G \cdot \Phi_G\rangle = \langle \theta_I|r_\sigma|\theta_G\rangle\langle \cdot \Phi_I|\Phi_G\rangle \qquad (3.13)$$

We can now consider the role of both the electronic and the nuclear part of this equation. The Raman process is so fast that despite the fact that energy is transferred to or away from the nuclei, no appreciable movement can occur during the time of any one scattering event. This means that the electronic part of the wave function can be approximated to what happens when the nuclei are at rest with a correction term to allow for the change in electronic structure when the nuclei move. To make this a little simpler, the electronic integral from the expression above is written as

$$\langle \theta_I|r_\sigma|\theta_G\rangle = M_{IG}(R) \qquad (3.14)$$

The movement is described by a Taylor series with the value at rest being the first and largest term $M_{IG}(R_0)$ where R_0 represents the co-ordinates at the equilibrium position. The second and higher terms describe the effect of movement along a particular co-ordinate R_ε and even the second term is relatively small. Thus all but the first and second terms can be neglected. For simplicity the first and second terms are written as M and M':

$$M_{IG}(R) = M_{IG}(R_0) + \left[\frac{\delta M_{IG}}{\delta R_\varepsilon}\right]_{R_0} R_\varepsilon + \text{higher order terms} \qquad (3.15)$$

In this way, the KHD expression can be solved. We will not attempt the mathematics here but they can be found in reference 6. Carrying out this procedure leads to the equation below. It looks complex but can easily be simplified.

$$(\alpha_{\rho\sigma})_{GF} = kM_{IG}^2(RO) \sum_I \frac{\langle\Phi_{R_F}|\Phi_{R_I}\rangle\langle\Phi_{R_I}|\Phi_{R_G}\rangle}{\omega_{GI} - \omega_L - i\Gamma_I} \quad \text{(A-term)}$$

$$+ kM_{IG}(RO)M'_{IG}(RO) \sum_I \frac{\langle\Phi_{R_F}|R_\varepsilon|\Phi_{R_I}\rangle\langle\Phi_{R_I}|\Phi_{R_G}\rangle + \langle\Phi_{R_F}|\Phi_{R_I}\rangle\langle\Phi_{R_I}|R_\varepsilon|\Phi_{R_G}\rangle}{\omega_{GI} - \omega_L - i\Gamma_I} \quad \text{(B-term)}$$

$$(3.16)$$

The two terms shown in the equation are known as A-term and B-term. Outside the summation sign, there is a term corresponding either to the electronic component of the Raman scattering (M) squared or to M times the much smaller correction factor M'. Thus, this part of the expression is much larger in A-term than in B-term. The summation sign ensures a contribution from all excited states to both A-term and B-term. However, as has been stated previously, the closer the excited state is to the laser frequency, the smaller the denominator in the first term of the equation and hence potentially the larger the contribution from the state.

In A-term, the numerator inside the summation sign consists simply of a multiplication of all possible vibrational wave functions. There is a theorem called the closure theorem which demonstrates that when all vibrational wave functions are multiplied together, the final answer is zero. Thus, no Raman scattering will be obtained from A-term. In B-term, an operator, the co-ordinate operator R_E, is present in the numerator. This operator describes the effect of movement along the molecular axis during the vibration and appears because the correction term M' has been multiplied out with the vibrational states. One feature of this operator is that the integral will only have a finite value when there is one quantum of energy difference between the initial state on which it operates and the final state. This means that only vibrations containing one quantum of energy will give Raman scattering. Thus, the theory will predict no overtones in Raman scattering and this is a good selection rule. Overtones are not seen unless there is some form of special effect. In addition, it is now possible to see how the Raman selection rule that symmetric vibrations are allowed comes about. The operators in the integrals in the numerator are dipole operators which, as for infrared absorption, are u in character. However, the Raman process requires that both integrals are multiplied out together. In essence this leads to a final result which is g in character.

3.10 LATTICE MODES

One type of vibrations which has not been considered so far are vibrations created in solid samples by radiation interacting with a lattice. For example, sodium chloride and silicon are materials which give vibrational spectra, but there is no definable molecule in which the atoms are linked by covalent bonds.

In this case, when radiation interacts with the material it induces vibrations through the whole lattice. One type of vibration forms along the direction of propagation of the radiation (longitudinal or L modes) and the other forms at right angles to it (transverse or T modes). These modes form through the whole crystal and each one consists of a very large number of vibrations of similar energy which occupy a band of energies in the material. The band breadth varies depending on the material. By studying bands of this type, as mentioned in Chapter 6, Raman spectroscopy can be used to study the properties of elements such as silicon. These bands are called lattice modes.

In sodium chloride, two types of lattice modes exist. In one type, the displacment is such that the chloride and sodium ions move together and in the other it is such that they move against each other causing a charge separation. The former type of lattice mode is lower in energy, with frequencies falling normally in the acoustic energy range. Modes of this type are called acoustic modes and labelled L_A and T_A. The higher energy type are called optic modes and labelled L_O and T_O. A full description of these modes is best given through a band theory approach and is outside the scope of this book. However, the term 'lattice mode' and the labels L_O and T_O are often used and the reader should be aware of them. In the case of silicon, the effect of loss of order in going from crystalline silicon to amorphous silicon is marked with the Raman spectrum broadening and shifting in frequency. This change is used in the electronics industry (see Chapter 6). In any study in which low frequency modes are important, the possibility of lattice modes being present should be considered.

3.11 CONCLUSIONS

The mathematics in Section 3.9 can be quite complex. This is perhaps not surprising since the equations have to describe the molecule in a distorted state at the instant that there is an interaction between the laser radiation and the molecule. However, some of the conclusions are quite simple and the analysis given above provides an insight into the background theory for Raman scattering which helps with problems and more detailed interpretation. Symmetry labels are commonly used in the literature, and at a basic level, the spectroscopist needs to understand their meaning to aid comprehension of many articles. However, the use of symmetry also improves understanding of the nature of vibrations and gives insight into the science that underlies the selection rules. The use of scattering theory is essential to understand the Raman process and such features as the weakness of overtones. It is also essential to understand resonance Raman scattering which forms the subject of Chapter 4. However, much of the rest of this book after Chapter 4 can be understood with only a minimum appreciation of the contents of this chapter.

REFERENCES

1. D. Long, *The Raman Effect: A Unified Treatment of the Theory of Raman Scattering by Molecules*, John Wiley & Sons, 1977.
2. J.R. Ferraro and K. Nakamoto, Introductory Raman Spectroscopy, Academic Press, San Diego 1994.
3. F.A. Cotton, *Chemical Applications of Group Theory*, Wiley Interscience, 1990.
4. R.J.H. Clark and T.J. Dines, *Angew. Chem., Int. Ed. Engl.*, **25**, 131 (1986).
5. R.J.H. Clark and T.J. Dines, *Mol. Phys.*, **45**, 1153 (1982).
6. D.L. Rousseau, J.M. Friedman and P.F. Williams, *Topics in Current Physics*, **2**, 203 (1979).

Chapter 4

Resonance Raman Scattering

4.1 INTRODUCTION

In the early days of Raman scattering, many spectroscopists preferred to avoid coloured compounds. After all, if a powerful beam of visible radiation is used to excite a molecule which is coloured, the light is liable to adsorb into the sample. This can cause strong fluorescence and prevent Raman detection. Even if it does not, it can cause sample decomposition through photodecomposition or heating. However, when the frequency of the laser beam is close to the frequency of an electronic transition, scattering enhancements of up to 10^6 have been observed and they are quite often of the order of 10^3 or 10^4. This means that Raman spectroscopy becomes a much more sensitive technique and since only the chromophore gives the more efficient scattering, it will also be selective for the part of the molecule involving the chromophore. When the resonance condition occurs, it turns out that it is possible to get electronic as well as vibrational information from the sample. One key reason this technique has become important is that the molecules give rise to good Raman scattering rather than intense fluorescence. They include the porphyrin rings which are present at the centre of a number of key enzymes, the pigments made from phthalocyanines, and other important classes of molecules such as the poly-acetylenes. For these species, resonance Raman scattering can give extremely informative *in situ* analysis and for this reason the use of resonance has been growing in recent years. The theory is well described in a number of reviews (see references [4–6] of Chapter 3). Here we will not carry out a rigorous mathematical treatment but concentrate on understanding the mechanism which underlie the equations used.

Modern Raman Spectroscopy – A Practical Approach   W.E. Smith and G. Dent
© 2005 John Wiley & Sons, Ltd   ISBNs: 0-471-49668-5 (HB); 0-471-49794-0 (PB)

4.2   THEORETICAL ASPECTS

4.2.1   The Basic Process

To obtain resonance Raman scattering, a laser beam is chosen which has an excitation frequency close to that of an electronic transition. Ideally, a tuneable laser would be used for excitation and the frequency would be chosen to correspond exactly to the energy difference between the ground vibrational state and the first or second vibronic state of the excited state. This condition is shown in Figure 4.1. Fortunately, the maximum resonance Raman scattering is not required for observing the effect or to obtain some enhancement. It is often more practicable to use an existing laser line available in the laboratory which has a frequency as near as possible to the true resonance frequency.

Figure 4.1 could also be used to explain the nature of the absorption process in electronic absorption spectroscopy. It shows a transition from the ground state to an excited state. The key difference, which cannot be seen in the diagram, is the length of time the molecule remains in an excited state. As we know from Chapter 3, the scattering process is fast, with scattering (the downward process in the diagram) occurring before the nuclei reach equilibrium positions in the excited state. In contrast, in absorption, the upward transition is also fast but the electron is absorbed into the molecule and the nuclei relax to the equilibrium geometry of the excited state. Thus, the processes of resonance Raman scattering and absorption are separated clearly by time, a variable not shown in the diagram.

Further, in the practice of absorption spectroscopy, the light irradiating the sample is usually polychromatic covering a wide range of frequencies, and

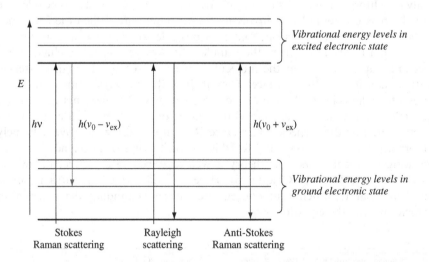

Figure 4.1.   Diagram of the basic process of resonance Raman scattering.

under these circumstances a number of transitions are involved. Often, the most intense transition is to one of the higher vibronic levels. The rounded shape of absorption bands is partly due to contributions from a number of levels and partly to the presence of hot bands arising from electrons present in excited states. However, as we shall discuss later in this chapter, the theory predicts that the most intense resonance Raman scattering in some cases will come mainly from the first two vibronic levels. As a result, it is not necessarily the case that the maximum absorbance of a UV visible transition is the energy at which the greatest resonance Raman scattering will be obtained from the excited state.

Clearly, with radiation of a frequency suitable to cause resonance, absorption as well as scattering will occur. When absorption occurs, the energy may be lost either by transfer to the lattice and dissipation as heat or as fluorescence. The ratio of scattering to absorption is a property of the molecule and is difficult to predict. From a practical point of view, fluorescence interference limits the number of molecules which will be suitable for examination by resonance Raman scattering.

The increase in intensity from resonance enhancement can be understood by studying the KHD equation analysed in Chapter 3 (Equation (3.11)). Here we will explain how the effect occurs with minimum reference to the mathematics. Full accounts of the mathematics and more in-depth references can be found in the references to Chapter 3. Consider the denominator of the first term. The resonance condition is met when the energy difference between the lowest vibrational state of the ground electronic state G and the resonant vibronic state I (ω_{GI} in Equation (3.11)) is of the same energy as the exciting light ω_L. This would mean that the denominator of the first term reduces to $i\Gamma_I$, which is a small correction factor. In some early forms of this equation it was not present, but it is required to take account of the lifetime of the excited state. Thus, under resonant conditions the denominator is very small and this will lead to the first term becoming very large, increasing polarizability and giving very much greater Raman scattering. Fortunately, in the second term in Equation (3.11), ω_{IF} and ω_L are additive and consequently this term can be neglected.

Another key difference between resonance Raman scattering and Raman scattering is immediately obvious from the KHD equation. In the resonance condition, almost all the interaction is with the one state, so that the Σ sign in the KHD equation can be dropped. This means that the scattering will depend on the properties of that state. As a result, the closure theorem, which states that the sum of all the vibrational matrix elements of a molecule is zero, is no longer valid. This was the reason that the A-term in Equation (3.16) did not predict any Raman scattering and since it is no longer valid, the A-term as well as the B-term can give resonance Raman scattering. This leads to two forms of resonance Raman scattering which have quite different properties.

For B-term scattering, as described in Chapter 3 (Equation (3.16)), the co-ordinate operator R_ε allows a transition only if there is one vibrational unit difference between the ground and the excited states. That is, no overtones were predicted. However, in A-term scattering there is no co-ordinate operator in the numerator. As a result, overtones in resonance Raman scattering, where it arises from A-term, will be allowed and there is no reason why they should not be intense. We shall see later some examples where there are intense overtones from A-term resonance Raman scattering. Further, in B-term scattering although the co-ordinate operator still exists, it is no longer a sum over a large number of states. As a result of this and possibly because of higher terms which result from a more complete analysis of the KHD equation, the overtone selection rule is not as effective in B-term resonance Raman scattering as in Raman scattering, and weaker overtones are obtained.

The interaction of the exciting radiation with the excited electronic states of the molecule is different in A-term and B-term enhancement. In A-term, the excitation which causes the scattering simply couples the ground state and the excited state as described previously. This type of scattering is called A-term or Franck Condon scattering. The electronic term M is much larger for scattering which arises from an A-term mechanism than from a B-term mechanism since it is M^2 as opposed to $M \times M'$ and A-term scattering might be expected to give more intense spectra than B-term (see Equation (3.16)). However this is only one factor. To obtain intense scattering, the transition should start from a point where there is considerable electron density in the ground state and go to a state where the wave function is such that, once populated, there will also be considerable electron density. Since the transition is vertical, this is often called good overlap between the states. In addition the selection rules must not prohibit the transition. This is similar to the conditions required for an intense allowed electronic transition in electronic spectroscopy. The matrix elements for the two processes are the same, which is not surprising, but it suggests that resonance enhancements will be most intense with allowed electronic transitions. In addition, strong A-term resonance enhancement occurs when there is a difference in the nuclear geometry between the ground and excited states.

In the B-term case, two excited states are mixed through the co-ordinate operator R_ε. Every time the molecule moves, and the geometry changes, there will be a need to remix the electronic states to obtain states new to the molecule. Thus, if there are two $\pi \rightarrow \pi^*$ transitions reasonably close together in the visible region, as is the case with porphyrins and phthalocyanines, then the co-ordinate operator will help to mix these two together. This type of more complex enhancement is called B-term or Herzberg Teller enhancement. One of the differences between A- and B-term enhancement is that B-term enhancement is only strong from the zero and first vibronic states of the excited state in resonance. There is no restriction on the excited vibronic states which can give A-term enhancement. In addition, B-term enhancement can arise from weak or forbidden transitions. For example the lower energy $\pi \rightarrow \pi^*$ transition of porphyrins is forbidden and weak, but B-term scattering from this transition is appreciable. This is because the orbital mixing process involves the higher energy $\pi \rightarrow \pi^*$ transition which is allowed.

4.2.2 Electronic Information

In Raman scattering, the KHD expression sums a number of small terms over all vibronic excited states of the molecule. As a result, it is difficult to obtain any electronic information from the spectrum. However, in resonance Raman scattering, one particular vibronic excited state is picked out as providing much of the enhanced scattering. As a result, the nature of the scattering depends on the nature of that state. It should be noted that other vibronic states close in energy to the one in resonance may also contribute, but it is easier to explain the effect by concentrating on one single state.

With A-term scattering, as stated above, no resonance enhancement will be obtained unless there is a change in the nuclear distances between the ground and the excited states. This actually means that small molecules such as iodine provide good A-term enhancement. In some of the older reviews it is stated that small molecules are the only molecules that give A-term enhancement. This is not completely true. Although large nuclear displacements between any two atoms are not usually obtained with large molecules, the sum of the many small displacements that occur can still give an appreciable nuclear displacement and therefore A-term enhancement.

To understand the effect that nuclear displacement has, it is necessary to consider the timescales of electronic and vibrational movement again. In essence, vibrational movement occurs over a longer timescale than Raman scattering. As a result, during the complete excitation and scattering process caused by any one photon, the molecule can be at any place along a vibrational co-ordinate. Since effective overlap between the ground state and the excited state in resonance is required to achieve the most efficient Raman scattering, it is necessary that this overlap is present through at least most of the vibrational cycle. A good example of this is in the resonance Raman spectrum of the molecule 5,5'-dithiobis-2-nitrobenzoic acid, and an ion which arises from it. This is a standard reagent used in electronic absorption spectroscopy to measure thiols in molecules since it provides coloured ions on reaction with thiols. The reactions are shown below. The compound, which was first used by Ellman, is often called Ellman's reagent, and here is simplified to ESSE.

$$ESSE + RSH = ESSR + ESH$$

$$PH\ 7.4$$

$$ESH = ES^- + H^+$$

The original molecule ESSE has an absorption band at 325 nm. However, the coloured ion ES^- formed at neutral pH has an absorption band at 410 nm. When a laser with an excitation wavelength of 457.9 nm (this was the closest wavelength to 410 nm available in the laboratory at the time) was used to excite ESSE, a typical Raman scattering spectrum was obtained with bands arising from vibrations of the phenyl rings (ϕ), the nitro group and the S–S bond identifiable. This gives an example of normal or perhaps pre-resonant spectra with most vibrations that would be expected to occur in the spectral region studied being present in the spectrum. However, the Raman scattering from ES$^-$ is much nearer to resonance and consequently much stronger. The spectrum of ESSE (Figure 4.2) was obtained from a 10^{-2} M solution, but to obtain an approximately equal intensity of scattering, the spectrum of ES$^-$ was obtained from a 10^{-5} M solution. Because ES$^-$ is near to resonance, the bands which are resonantly enhanced depend on the excited state in resonance. The spectrum consists of only two main peaks which are assigned to the nitro group stretch and bend. On the assumption that this is A-term scattering, to get effective resonance enhancement a nuclear displacement and good overlap are required. The biggest change in nuclear displacement of this molecule will be along the nitro group bonds. The high intensity of the stretching motion along these bonds suggests that the excited state has an electronic geometry which is elongated along the bond thus giving good overlap at any point on the main stretching vibration. This experiment indicates the selectivity of resonance Raman scattering and also one of the ways in which electronic information can be obtained.

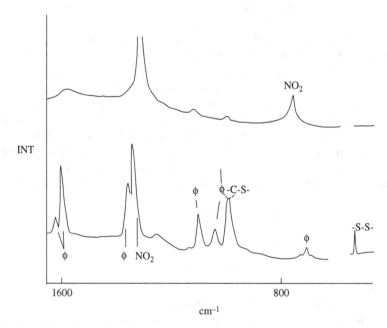

Figure 4.2. The spectrum of Ellman's reagent at 10^{-2} M and of the anion ES$^-$ at 10^{-5} M.

Another clear difference between normal Raman scattering and resonance Raman scattering is that in both cases the intensity is dependent on the fourth power of the frequency, but in resonance Raman scattering the intensity is also strongly dependent on how close the frequency of the excitation used is to that of the frequency of an allowed electronic transition. It is worthwhile to look at the KHD expression once again. As previously discussed, in the resonance condition the denominator is very small. However, as the difference between the laser excitation frequency and the transition frequency increases, the resonance enhancement will drop off quite quickly. If the total intensity obtained in resonance is given the value 1, then 10 wavenumbers away from resonance the enhancement will be down to about a tenth and at a 100 wavenumbers about a hundredth, etc. Thus, we need to be quite close to resonance to get the maximum enhancement. Perhaps of more importance and not widely recognized by many Raman spectroscopists is what happens far away from resonance. Supposing for example the total enhancement obtained was 10^4, then 10,000 wavenumbers away from resonance, an enhancement of one would be obtained. Thus, there is a very long tail on the frequency dependence of the resonance enhancement process and it could be that even with infrared excitation, enhancement factors of 5 or greater could be obtained for molecules with visible chromophores. This may seem trivial compared to the enhancements that can be obtained close to resonance but it does mean that a coloured part of a molecule could easily have a spectrum five times as intense as would have been expected in an ordinary Raman spectrum even if an FT infrared system is used. If the molecule is a strong Raman scatterer this can mean that a minor component can be picked out selectively. We will show later in this book (Chapter 6) that this can be used to good effect, for example in looking at ink jet dyes with Raman scattering obtained with an infrared laser.

The above explanation is over simplified. It should be noted that for any one process a number of vibronic bands which are near but not quite in resonance would contribute and this will change the frequency/intensity relationship. The greater the separation between the laser frequency and the frequency of the resonant excited state, the more appreciable the contributions from other vibronic states is to be expected until normal Raman scattering is essentially restored.

4.2.3 Resonance Excitation Profiles

Since the absolute and relative intensities of the bands in a resonant spectrum are dependent on the separation between the excitation and resonance frequencies and on the nature of the electronic states, it can be useful to plot the intensity of selected bands against the frequency of the laser. This is called a resonance excitation profile (REP) and to do this effectively, a tuneable laser is preferred so that the Raman spectrum can be recorded at a large number of different laser frequencies. An approximation to this can sometimes be achieved by using a multiline laser if the absorption band is sufficiently broad and there are enough lines from the laser available in the energy region covered by the absorption band. In the simplest REP, the maximum intensity of the band produced will be the point at which resonance occurs. If the vibration plotted is resonant with more than one vibronic level, more than one peak should be seen in the profile. Also, since this is resonance and not all vibrations are affected in the same way, the profiles will vary from one vibration to another. This information can be difficult to obtain and requires quite sophisticated equipment. However, it provides unique and extremely valuable information to probe in depth the electronic and vibrational structure of a particular molecule. An example of a resonant excitation profile for two different vibrations is shown for the phthalocyanine molecule in Figure 4.3. One, ν_3, is at a relatively high frequency and the structure on the profile indicates that a number of vibronic states are involved. This is reinforced by the upward slope to higher frequencies and suggests some A-term enhancement. The lower frequency vibration, ν_7, shows two main bands more typical of B-term enhancement. A weaker peak can also be seen on the low energy side of the main band in the ν_7 profile. This is because the excited state of the phthalocyanine is split

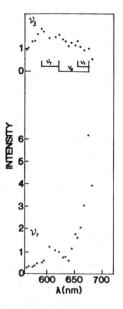

Figure 4.3. Resonant excitation profiles for two copper phthalocyanine vibrations $-\nu_3$ and ν_7 created using different excitation wavelengths.

by a dynamic Jahn Teller effect and the extra weak band arises from a profile due to a second electronic state.

The above description provides only a qualitative understanding of the nature of resonance Raman scattering. A more in-depth review can be obtained in the references to Chapter 3. However, enough has been said to make practical use of resonance Raman scattering and also to interpret the basic effects seen in the spectra. Table 4.1 gives the main similarities and differences between Raman scattering and resonance Raman scattering.

Table 4.1. Main differences between Raman scattering and resonance Raman scattering

Raman scattering	Resonance Raman scattering
B-term effective	A- and B-term effective
No overtones	Overtones common
More modes observed in the spectrum	Some modes selectively enhanced
No electronic information	Electronic information present
Weak scattering	Stronger scattering

4.3 PRACTICAL ASPECTS

As already discussed, in resonance Raman scattering the laser energy is chosen to match as closely as possible that of an absorption band of the analyte. As a result, absorption will occur which may cause both sample decomposition and fluorescence depending on the nature of the material. Therefore, the spectroscopist must develop a strategy which minimizes these effects. To assess the extent of the problem, one obvious thing to do is to observe the sample before and after exposure to the laser beam. Very often with resonance Raman scattering, sample damage can be clearly seen in a coloured sample as a change in colour or a black spot. However, in some cases, such as with proteins containing the heme group, lower powered radiation alters the protein structure but does not destroy the heme and hence there is little to no visible change in colour. It is important that the spectroscopist recognizes these subtle changes. In principle, it would be easy to check this using absorption spectroscopy. However, the difference in sample volume actually interrogated in Raman scattering and in absorption spectroscopy can give misleading answers. In most absorption spectrometers, the sample is placed in a 1 cm cuvette and the beam passes through a significant part of the sample volume whereas if Raman scattering is obtained from the same cuvette, it is usually obtained from a small focussed volume. Damage, which occurs in this volume in a short time, will affect the Raman scattering, but may not be sufficient to be observable in the absorption spectrum of the bulk of the sample.

One way to minimize photodegradation is to use a sampling method in which the sample passes through the laser beam but does not stay in the beam for the whole time of the analysis. This way any one part of the sample is not retained in the beam for any lengthy period of time. Raman scattering is then obtained from the cumulated spectra from a large area of the sample. For example, with solid samples, spinning disks are often used, as already mentioned in Chapter 2, Section 5. In this technique, the sample is pressed into a disk, or compressed into a channel cut in a blackened support disk. The laser is then set to focus on the outer part of the disk or onto the channel in the disk and the disk spun. In this way, scattering is collected from the point at which the beam is focussed but the sample precesses through the beam limiting the exposure time of any one area of the sample. This process allows excited states to relax and heat to diffuse away from the sample before the disk completes a revolution and the same part of the sample is interrogated again. This is effective, although in many samples, the track where the sample has decomposed can clearly be seen. It still means there is less decomposition than would be present in a statically focussed sample.

Similar devices can be devised for use with solution samples (Chapter 2, Section 5). They usually consist either of a spinning sample holder, such as an NMR tube with the beam tightly focussed close to the surface, or a small flowcell where the sample can either flow through or oscillate back and forward under the laser beam.

The spectroscopist has another approach which may be effective. As discussed earlier in this chapter, the resonance contribution extends over a range of frequencies decreasing as the difference between the frequency of the absorption band and the excitation laser increases. Thus, by moving the frequency of excitation away from the resonance frequency to a pre-resonance frequency, it is possible to avoid the worst effects caused by absorption of the excitation while still retaining a degree of enhancement. With pre-resonance excitation as discussed above, it should be borne in mind that the further away from resonance the spectrum is recorded, the more normal Raman scattering selection rules will apply. In addition, with molecules of high symmetry, the different symmetry types of vibration have different dependencies of resonance enhancement on frequency.

This type of experiment highlights another possible problem with resonance. In a non-resonant sample, a visible laser beam can be focussed tightly within the media and although there are refraction and reflection effects, they are minor compared to those obtained on focussing into a coloured solution to obtain resonance. In normal Raman scattering, the focussed spot created by the laser beam can be effectively imaged back onto the detector. However, in the resonant condition, the laser light is absorbed by the medium and the deeper it penetrates, the less intense is the light. Thus, the focussed spot may occur at a position in the sample where there is relatively little laser power. Further, the scattered radiation is much weaker than the exciting radiation and is given out as a cone. Therefore, it will be absorbed rapidly as it passes back through the sample. This effect is called 'self-absorption' and can make resonance Raman scattering difficult to obtain despite the enhancement.

In solution, there is usually a concentration range in which resonance scattering is effective. If the sample is too concentrated, the beam may be so attenuated as to be weak at the focus spot and in this condition, self-absorption of the scattered radiation will prevent effective collection of the Raman scattering. In addition, where there is beam penetration into the sample, local heating caused by absorption can cause a change in the dielectric constant of the medium, and cause a lensing effect along the beam. This makes it much more difficult to collect the light effectively. However, if the sample is too dilute, the Raman scattering will be too weak to detect. Thus, it is important while measuring solution resonance Raman scattering to recognise that if poor scattering is obtained, it may be necessary to dilute the sample rather than increase the concentration. It is possible to find the right concentration only by trial and error.

To minimize these effects, the beam should be focussed close to the sample surface, but there is a limit as to how well this can be done. In a solid, specular reflection from the surface can occur if the beam is focussed directly on the surface, and in solution, focussing too close to the glass wall at the front of the sample can cause intense reflection from the glass. The sudden appearance of more intense light under these conditions usually means that the laser is focussed on the glass rather than on the solution. This apparently intense scattering can mislead the spectroscopist into recording spectra of the wall of the vessel rather than the solution. The ready availability of disposable plastic cuvettes has made this a more serious problem. When focussing through them, the spectrum of the plastic is not usually observed, but if the beam is focussed on the wall of the cuvette, excellent spectra can be recorded from the polymer material which can easily be mistaken for the spectrum from the sample.

Despite these problems some key results can be obtained from resonance Raman scattering which make it an important technique in some fields. Some examples of where it is effective are given in Section 4.4.

4.4 EXAMPLES OF THE USE OF RESONANCE RAMAN SCATTERING

4.4.1 Small Molecules

The Raman spectra of iodine is a classic example of A-term resonance enhancement from a small molecule. Excitation of I_2 vapour obtained by heating some iodine in a vapour cell produces a remarkable spectrum consisting of a series of sharp lines which are very intense (see reference 6 of Chapter 3). As discussed in Chapters 1 and 3, we expect one vibration from a diatomic molecule in the gas phase and the frequency of the lowest energy peak corresponds approximately to the energy we would expect from Hooke's law. The other peaks appear at regular intervals and the energy between them is approximately one quantum of the same energy as that of the vibration. A similar experiment is easily performed either by using a solution of iodine or by focussing onto iodine as a solid. Figure 4.4 shows the spectrum of solid iodine recorded from an iodine crystal using a Raman microscope and 514.5 nm excitation. Again there is a regular pattern of bands but the bands here are broader due to solid state effects.

We are able to learn more about the nature of iodine from this spectrum than we would from conventional Raman scattering. First, the nearly equally spaced bands are due to the fundamental and overtones of the one vibration. The fundamental is the lowest energy peak and some overtones are more intense than the fundamental! This is clearly A-term scattering in which there is no selection rule to forbid the overtones occurring. Secondly, with the harmonic approach, we would expect the overtones to be equally separated in energy. However, as discussed in Chapter 3, with the true Morse curve, the separation decreases towards higher energies as a result of the non-harmonic nature of the curve. Thus, by studying the changes in separation, it is possible to calculate the shape of the Morse curve. This is another example of Raman scattering providing electronic information.

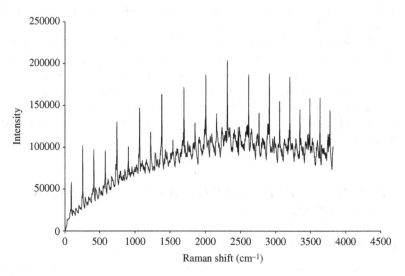

Figure 4.4. The resonance Raman spectrum for solid iodine taken with 514.5 nm excitation.

A second example of A-term scattering is given in Figure 4.5 for the pigment lapis lazuli. A synthetic form of this is widely used as the pigment ultramarine. For many years, the colour in this compound was a mystery. However, it is now known to be due to the small sulphur species, S_3^- and S_2^-, which are trapped in an oxide lattice. It is the transitions from these ions and particularly S_3^-, which give the colour in the visible region. Figure 4.5 shows the resonance Raman scattering taken from a sample of ultramarine using 406 nm radiation for excitation. This wavelength will enhance the intensity of S_3^- over S_2^- since it is closer to resonance with it. S_3^- will have more than one possible vibration, which could be observed in the Raman scattering. In essence, a symmetric stretch and a bend are the most likely to appear. The band at about $500 \, \text{cm}^{-1}$ is the fundamental mode of the stretch. The fundamental of the bend is closer to the exciting line and not observed because of the filter used. The bands at higher energy are overtones and combination modes of the two fundamentals. The most intense fundamental band is the stretch and the most intense peaks with regular separations are the overtones of it. Again, there is a slight decrease in the frequency of the separations between the bands for higher overtones and this information can be used to calculate the nature of the Morse curve for the ground state of ultramarine. The weak band just above the fundamental of the stretch is a combination mode corresponding to one quantum of the stretch and one of the bend. Further, by using a different frequency (normally 457.9 nm radiation from an argon ion laser) it is possible to ensure a greater enhancement of the peaks due to S_2^- and consequently to estimate *in situ* the approximate ratio of S_2^- to S_3^- from resonance Raman scattering.

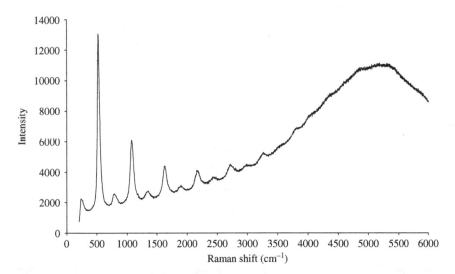

Figure 4.5. The resonance Raman scattering taken from a sample of ultramarine using 406 nm excitation. The broad underlying band is fluorescence.

4.4.2   Larger Molecules

Perhaps the most widespread use of resonance Raman scattering is for the study of heme-containing proteins and there are good reviews on this topic [1, 2]. The resonance Raman scattering obtained with visible radiation from these proteins is due to an interaction with the $\pi \rightarrow \pi^*$ transitions from the porphyrin ring of the heme group (Figure 4.6).

Detailed assignments of the vibrations have been made and some of these vibrations which appear strongly in resonance Raman scattering can be linked to structural properties. For example, the top three vibrations shown in Figure 4.6 are used as markers for the oxidation state and spin state of the ion. The

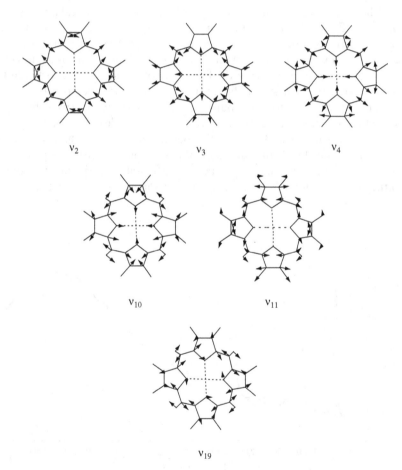

v_2 v_3 v_4

v_{10} v_{11}

v_{19}

Figure 4.6.   Typical vibrations taken from a porphyrin ring system (the symbol printed below each of the diagrams are the conventional ones used to describe that particular vibration).

oxidation state marker, v_4, shifts more in frequency than other bands when the oxidation state of the iron ion in the centre of the ring is changed. The reason for this can easily be seen. One of the largest displacements is on the four nitrogens attached to the central metal ion. They move symmetrically inwards and outwards in phase to alter the size of the hole in the centre. Thus, the size of the metal ion which fills this hole and which changes as the oxidation state changes would be expected to have a significant effect on the frequency of this vibration. The reason why v_3 and v_{10} act as spin state markers of the iron is more subtle. Their common feature is that they have significant displacements on the inner ring system of the porphyrin. There is much more information to be obtained from these spectra. For example, in protoporphyrin 1X there are two vinyl groups attached to the porphyrin ring system which are quite often not in the plane of the porphyrin ring. When this is the case there is only very weak conjugation between the vinyl groups and the heme system, and hence the intensity of the vinyl groups is low since there is little resonance enhancement. However, when they become planar with the ring, the conjugation increases and the band intensities increase. Thus, bands assigned to the vinyl groups can be important in deciding the position of these groups relative to the heme system. This alters with some protein-functional changes. Other vibrations which give effective Raman scattering can be related to the amount of doming or ruffling on the porphyrin ring.

The heme group has two sets of electronic absorption bands in the visible region as shown in Figure 4.7. The most intense band is the Soret band at about 410–450 nm and there are weaker bands between 500 and 650 nm called the $\alpha 1$ and β or Q_0 and Q_1 bands.

Excitation at the frequency of the Soret absorption band produces Raman scattering in which the A_{1g} vibrations are prominent. This might be expected, since if the heme ring is regarded as flat and peripheral groups such as the vinyl groups are ignored, it belongs to the D_{4h} point group. Thus, with an intense

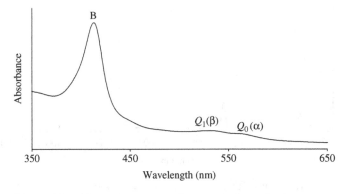

Figure 4.7. Absorption spectra for a typical porphyrin system.

allowed band such as the Soret band we would expect A-term scattering and hence the totally symmetric modes should be enhanced. However, other bands that appear in the spectra suggest there is also some B-term scattering.

When in resonance with the α1 and α2 bands, it is the B_{1g} and B_{2g} modes which tend to be the most enhanced, although A_{1g} modes can often be observed. We would expect this change of emphasis. B-term enhancement uses the Hertzberg Teller mechanism which was discussed very briefly at the end of 4.2.1 for this case. The co-ordinate operator mixes two excited states. Here the forbidden $\pi \rightarrow \pi^*$ transition which gives rise to the weak α1 and β absorption bands is in resonance with the laser. The states mixed are those which give rise to these bands and the Soret band. Effective mixing of the states requires a less symmetric vibration and hence the enhancement of the B_{1g} and B_{2g} bands. Figure 4.8 shows the resonance Raman scattering taken with 406 nm excitation from a P450 protein. It is expected in this spectrum that the most symmetric bands will dominate but the other bands will also be present. The most intense band is v_4, the oxidation state marker, at 1372 cm^{-1}. This energy position is characteristic of iron in oxidation state III. The bands v_{10} and v_3 are weaker but they are at energy positions which clearly indicate that this protein is in the high spin form. A more in-depth analysis of the complex spectra obtained in the 1600–1550 cm^{-1} region makes it clear that there are other changes occurring. These refer to the vinyl groups and the ring doming/ruffling described earlier. To analyse the spectrum which is shown in Figure 4.9, curve fitting procedures were used. The dangers of this approach were discussed in Chapter 2. In this

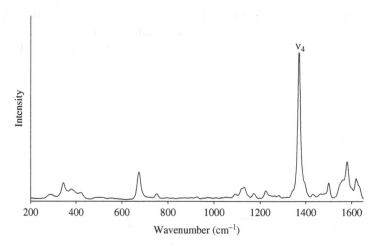

Figure 4.8. The resonance Raman scattering from a P450 protein taken with 406 nm excitation.

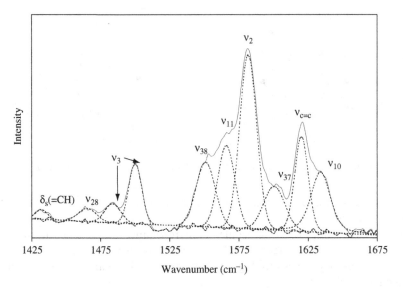

Figure 4.9. Spectra in the region from 1500 to 1650 cm^{-1} taken with 406 nm excitation and curve fitted. (Reproduced with permission from W.E. Smith and S.J. Smith, *Biopolymers*, **70**, 620–627 (2003).)

case, there is a wealth of related literature which makes some but not all the assignments definite but data of this type is best used with other evidence.

P450 proteins are large with a complex structure. Despite the very large number of atoms and bonds in this molecule the resonance spectrum picks out clearly the bands due to the heme to give a very good selective spectrum. However, it raises an interesting question as to what has happened to the normal Raman scattering from the rest of the protein. After all, given the large number of atoms which will contribute to non-resonant but Raman active vibrations, it would be expected that the summation of all vibrations of a particular molecule might appear as a broadened band somewhere in the spectrum. In fact, normal Raman scattering from a protein is weak but can be obtained. In the presence of strong resonance Raman scattering it is seldom seen.

This raises a general question about Raman scattering. Some types of vibration such as the vibrations of water are very weak and other vibrations even with normal Raman scattering can be quite strong. Thus, Raman scattering has a natural selectivity which here has been enhanced by using bands which would normally be quite strong in normal Raman scattering, under resonant conditions. The Raman spectroscopist really needs to consider this when considering the use of Raman scattering for a particular problem. If the species to be detected gives strong Raman scattering, either because the normal Raman scattering is strong, or because it is in resonance, it may well be a very effective technique. However, if the Raman scattering from the matrix is strong and

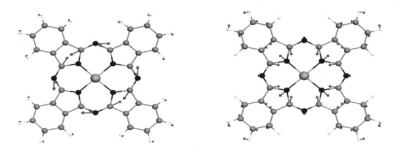

Figure 4.10.   Two characteristic vibrations of copper phthalocyanines.

from the substance which it is desired to detect it is weak, it may be difficult to obtain an effective spectrum.

Many other important molecules give resonance Raman scattering. One widely used class of pigments which give excellent resonance Raman scattering are the phthalocyanines. In Figure 4.10, two of the many vibrations in the molecule are shown.

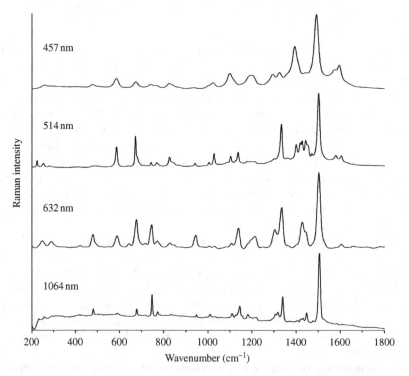

Figure 4.11.   The spectrum of copper phthalocyanine taken with four different excitation frequencies.

It should be noticed that there are strong similarities between these vibrations and the porphyrin system. This is perhaps not surprising since both contain ring systems which in their simplest form are of D_{4h} symmetry. On an initial glance, the Raman spectra of phthalocyanines can appear quite similar. For example, the frequency of the band assigned at v_3 here is affected by the size of the metal ion. For many metal ions the frequency is linearly dependent on size but for particularly big metal ions, the frequency is lower. This is believed to be because the largest ions can no longer fit into the plane of the ring and exist above or below the plane, allowing displacement to occur more easily. The NIR FT Raman spectra showing this effect are discussed in Section 5 of Chapter 6. The spectra show the band shift when metal-free, copper, gallium and titanium oxy-phthalocyanines were examined.

In Figure 4.11, the effect of changing the excitation frequency on the resonance Raman scattering from copper phthalocyanines can clearly be seen. In this case, and in contrast to the heme system, there are two allowed bands in the visible region, one in the blue and one in the red. There are also weaker bands which could give B-term enhancement by mixing with an allowed band. The band v_3 dominates the spectrum at all frequencies, even in the near infrared. However, if the spectra are inspected more closely, it is clear that each spectrum is different due to selective enhancement from resonance with different electronic states.

Finally, the overtones, which were so strong with A-term scattering from small molecules, also occur with heme and phthalocyanine systems. However, despite the presence of some A-term enhancement, they are much weaker, reflecting the smaller displacements of the nuclei during the vibration. Figure 4.12

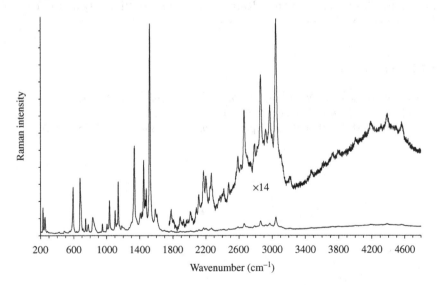

Figure 4.12. Overtones of copper phthalocyanine taken with an excitation frequency of 514 nm.

shows this for copper phthalocyanine. The most intense band corresponds to two quanta of the most intense fundamental band which is assigned as v_3 and has A_{1g} symmetry.

4.5 CONCLUSIONS

Resonance Raman scattering will not be effective for all molecules which are coloured. Much depends on the relative efficiency of the scattering and fluorescence processes. Fluorescence can easily dominate the spectra making it very difficult to obtain the Raman scatter experimentally. Sample decomposition and difficulties with self-absorption are also problems. However, for many systems the scattering is strong and there are strategies available to get round the worst interferences. These include the use of pre-resonance to avoid the worst problems with fluorescence and the use of spinning sample holders or flow cells to reduce photodegradation. There are key advantages in using resonance Raman scattering which make it worthwhile. It provides more intense spectra and consequently can be used to selectively pick out and positively identify a molecule in a matrix. Electronic information about a molecule, can be obtained from the intensities of the bands found in resonance, from the energy separations in overtone progressions, and from the overtone patterns that can be obtained. The weak nature of ordinary Raman scattering from molecules such as proteins and in particular water, make it possible to examine resonance Raman scattering directly in the presence of some other materials. This makes resonance Raman scattering a particularly useful form of Raman spectroscopy in biological systems.

REFERENCES

1. T.G. Spiro and X.-Y. Li in: *Biological Applications of Raman Spectroscopy*, T.G. Spiro (ed.), vol. 3, Wiley, New York, 1988, p. 1.
2. S.Z. Hu, K.M. Smith and T.G. Spiro, *J. Am. Chem. Soc.*, **118**, 12638 (1996).

Chapter 5

Surface-Enhanced Raman Scattering and Surface-Enhanced Resonance Raman Scattering

5.1 INTRODUCTION

Raman microscopists can obtain effective Raman scattering from very small amounts of solid material. Consequently Raman spectroscopy can be said to be sensitive in some circumstances for some solid materials, but for many applications such as analysis in solution, lack of sensitivity remains one of the key limitations. Thus, a Raman technique which could offer a major improvement in sensitivity would be valuable. Surface-enhanced Raman scattering (SERS) gives an enhancement of up to about 10^6 in scattering efficiency over normal Raman scattering, but its potential has until now been largely unfulfilled because of the complexity of the technique and problems with understanding the theory. Other techniques such as resonance Raman scattering are better understood and do provide a significant improvement in sensitivity but all have severe limitations. For example, with resonance Raman scattering, the technique is effective only for some coloured molecules and the problems of fluorescence interference and sample photodegradation can limit its use still further. The fact that SERS is effective with a wider range of molecules and gives a bigger enhancement in sensitivity makes the technique worth considering for some targets.

Recently, an increasing amount of work has been published on the use of SERS with analytes which have resonant chromophores. This technique is called surface-enhanced resonance Raman scattering (SERRS) and it has

Modern Raman Spectroscopy – A Practical Approach   W.E. Smith and G. Dent
© 2005 John Wiley & Sons, Ltd   ISBNs: 0-471-49668-5 (HB); 0-471-49794-0 (PB)

several advantages over SERS. Compared to SERS, it shows a considerable increase in enhancement, and has the sensitivity to rival or surpass that achievable with fluorescence. When a molecule with a chromophore which fluoresces is adsorbed on the SERS active metal surface, the fluorescence is almost completely quenched. As a result, the technique applies to a wider range of molecules than resonance Raman scattering or fluorescence, and molecules normally considered to be fluorophores can be detected by SERRS. Further, the *in situ* identification of a particular molecule from particular features in a spectrum is often much more certain with SERRS. For these reasons, SERRS, though underdeveloped, is now of potential interest as a probe to provide informative detection in areas such as bioanalysis and nanotechnology.

SERS was initially observed in 1974 by Fleischman *et al.* [1]. They reported strong Raman scattering from pyridine adsorbed from an aqueous solution onto a silver electrode roughened by means of successive oxidation–reduction cycles. The authors attributed the effect to a large increase in the electrode surface area caused by the roughening process which enabled more pyridine molecules to be absorbed on the surface. However, Jeanmarie and Van Duyne [2] and Albrecht and Creighton [3] showed that the intensity was due to more than the increase in surface area. They noted that the likely increase in intensity from the roughening of the surface would be less than a factor of 10, whereas the enhancement obtained was of the order of 10^6.

The basic technique consists of using an electrochemical cell into which is placed a solution of the analyte. A possible cell for use in these experiments is shown in Figure 5.1 (bottom). This cell is designed to fit into a glass cuvette and to position the working silver electrode in such a way that the scattered radiation from it is easily focussed into the spectrometer. It is not a particularly efficient electrochemical cell due to the position of the secondary electrode being dictated by the desire to leave the scattered radiation path unobstructed. There are many such designs which in different ways accommodate the conflicting demands of electrochemistry and radiation collection. The spectrum of pyridine taken with different voltages applied to the cell is shown in Figure 5.1 (top). The magnitude of the surface enhancement and the relative intensities of the peaks change depending on the voltage applied.

It has been demonstrated that silver is a particularly good substrate for SERS but some other metals are also effective. Gold is widely used and copper is known to give good enhancement. Other metals including lithium and sodium have also been shown to work well. Many different roughened surfaces have been prepared, the most common being aggregated colloidal suspensions, electrodes and cold deposited metal films including silver island films and silver coated beads.

A number of requirements need to be met for the surface to be SERS-active. It is essential that there is effective surface adsorption of the analyte. Often this is neglected and as a result, poor, variable results are obtained. In addition,

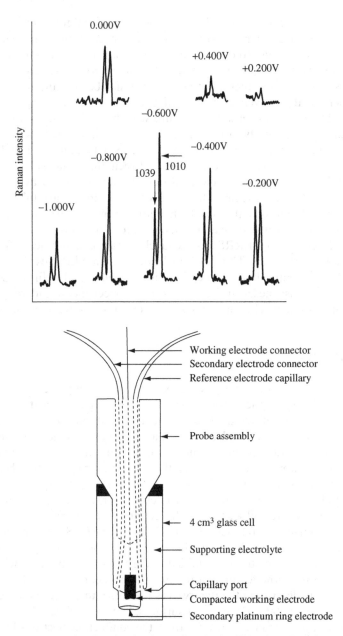

Figure 5.1. An example of a simple electrochemical cell for SERS and the spectra of pyridine at various potentials given as relative to SCE. Peak positions are in cm^{-1}.

SERS experiments can be carried out in a wide range of environments. These can vary from an atmosphere-controlled vacuum chamber containing small amounts of pure analyte to cuvettes open to the air and containing biological media as well as the analyte. It is possible to make a rough surface of iron and get some surface enhancement, but usually iron has been found to be ineffective. Copper metal gives very effective SERRS in some reported studies and not in others. The common reason for failure with both these metals is that the roughened surfaces rapidly form multilayers of oxide in the presence of oxygen. This alters the nature of the surface and can lead to annealing and loss of surface roughness. Further, the distance between the metal surface and the outside of the oxide layer can act as a barrier spacing the analyte away from the actual silver metal and reducing SERS. Only a few authors have fully described the chemical precautions they took in defining the surfaces. The effect of bubbling oxygen or nitrogen gas through solutions has been shown to have a considerable effect on SERRS due to a chemical or a physical effect or a combination of the two. Thus, before beginning an SERS experiment it is essential to make sure that the chemistry is controlled and defined. Once this is ensured, it is also necessary to choose the correct metal and roughen the surface appropriately. The following section explains the reasons for this.

5.2 THEORY

Since this technique was discovered experimentally, many theories were proposed, particularly in the early stages. To some extent almost all of them contain an element of truth. The problem is that our ability to describe theoretically the bonding or adsorption of an organic molecule to a roughened, probably corroded and oxidized metal surface in water is very limited. In essence most authors now accept that there are two parts to the theory and that both have some validity. This provides a working theory which enables experimental work to proceed with little controversy over the nature of the effect. However, as we will see later in this chapter, the true nature of the theory is still a very active research field. It is likely that the SERS effect is due to one single cause and possible that once the adsorption/complexing process on the surface is better understood, a unified theory will be obtained which will incorporate most of the features of the two theories currently used.

Before briefly describing the theory, it is necessary to understand the nature of the roughened metal surface. Silver surfaces, like the surfaces of other metals, are covered with electrons. They arise from the conduction electrons held in the lattice by the presence of positive charge from the silver metal centres. At the surface, the positive charge is only on the metal side of the electrons. Consequently the electron density extends a considerable distance from the surface and there is also freedom of movement in a lateral direction along it. When a light beam interacts with these electrons, they begin to oscillate as a collective group across the surface. These oscillations are termed surface plasmons.

Surface plasmons from small uniform particles, or from surfaces which have a single periodic roughness feature, have a resonance frequency at which they absorb and scatter light most efficiently. The frequency varies with the metal and the nature of the surface. It so happens that both silver and gold plasmons oscillate at frequencies in the visible region and therefore, they are suitable for use with the visible and NIR laser systems commonly used in Raman scattering. On a smooth surface the oscillation occurs along the plane of the surface. Absorption can occur but no light will be scattered. To get scattering, there needs to be an oscillation perpendicular to the surface plane and this is achieved by roughening the surface. This locates the plasmon in the valleys of the roughened metal surface and scattering is caused as the plasmons move up to the peaks. These peaks from which scattering is deemed to occur are sometimes called 'lightning rods'.

Other features of the metal are vital. First, metals can both scatter and absorb but the ratio of the two is metal-dependent and, compared to other metals, the ratio for silver favours scattering. To a physicist, the dielectric constant of the metal is divided into two parts, the real and imaginary. Scattering is associated with the real part and absorption with the imaginary part. Some of the SERRS papers use this terminology but it is largely beyond the scope of this book; however, further information can be found in reference [4].

In addition to the ratio of absorption to scattering, the nature of the roughness is important. Usually for a metal surface obtained either by electrochemical roughening of the electrode or by depositing silver onto a surface, there are many different roughness features of varying dimensions. The result of this is that the plasmon on the surface usually covers quite a broad range of wavelengths. It is simple to determine this by measuring the absorption spectrum of the plasmon. However, with nearly mono-dispersed colloids, the range of frequencies covered by the absorption band is very much narrower and usually the half-width is about 50–60 nm indicating a much more defined surface roughness.

Thus, to obtain good SERRS, the surface must be reasonably clean or at least must not form too thick an oxide layer, and it must be suitably roughened in a way that is time-stable. In addition, it needs to be of a material that has plasmons which resonate in a frequency range which includes that of the exciting laser. The basics of how this is believed to give effective SERS is given in Section 5.3.

5.3 ELECTROMAGNETIC AND CHARGE TRANSFER ENHANCEMENT

There are two different theories of surface enhancement which are currently used [4–7]. In one, the analyte is adsorbed onto or is held in close proximity to the metal surface, and an interaction occurs between the analyte and the plasmons. This is called electromagnetic enhancement. In the other, the adsorbate chemically bonds to the surface. Excitation is then through transfer of electrons from the metal to the molecule and back to the metal again. This is called charge transfer or chemical enhancement. By definition it can only be possible from the first layer of the analyte attached to the surface whereas electromagnetic enhancement could occur from a second or subsequent layer. Some enhancement has been claimed up to about 20 Å or more away from the surface.

5.3.1   Electromagnetic Enhancement

The simplest description of electromagnetic SERS is based on models of a small metallic sphere. This is clearly not going to be adequate since there is no roughness of the type found to be important experimentally. For example, the simple sphere could be regarded as a first approximation to a single colloidal particle but we know that aggregation of suspensions of these particles gives much increased SERS. However the sphere model can be used to explain much of the basic process. The effect of aggregation can be dealt with after consideration of the sphere model.

When a small metal sphere is subjected to an applied electric field from the laser, the field at the surface is described by

$$E_r = E_0 \cos\theta + g\left(\frac{a^3}{r^3}\right)E_0 \cos\theta \qquad (5.1)$$

E_r is the total electric field at a distance r from the sphere surface,
a is the radius of the sphere,
θ is the angle relative to the direction of the electric field,
g is a constant related to the dielectric constants such that,

$$g = \left(\frac{\varepsilon_1(\nu_L) - \varepsilon_0}{\varepsilon_1(\nu_L) + 2\varepsilon_0}\right) \qquad (5.2)$$

ε_0 and ε_1 are the dielectric constants of the medium surrounding the sphere and of the metal sphere respectively. ν_L is the frequency of the incident radiation.

At some point where the denominator is at a minimum, the value of g will be a maximum. ε_0 is usually close to 1 and consequently this maximum usually occurs when ε_1 is equal to -2. At this frequency, the plasmon resonance frequency, the excitation of the surface plasmon greatly increases the local field experienced by the molecule absorbed on the metal surface. In essence, the molecule is bathed in a very freely moving electron cloud and that movement intensifies the polarization of the surface electrons. The electrons in the analyte molecule adsorbed on the surface interact with this cloud causing greater polarization around the molecule.

At the metal surface the total electric field is averaged over the entire surface of the small sphere. At any point on the surface the electric field may be described by two components, the average field perpendicular to the surface and the average field parallel to the surface. Clearly, g is dependent on the dielectric constants of the metal and the surrounding medium and also the laser frequency. Since the dielectric constant of the metal is generally about 1, it can be seen by substituting this into Equations (5.1) and (5.2), that the electric field is greater perpendicular to the surface than parallel to it. Thus, the greatest enhancement is observed for a molecule adsorbed on the surface and polarized perpendicular to it. Further, since the field is inversely proportional to r^3, the magnitude of the SERS enhancement drops off rapidly with distance from the surface.

It is now known from many experiments, mainly on small particles adsorbed on a surface, that the greatest enhancement does not occur evenly round every isolated particle but at points between some touching particles or clusters of particles. Enhancement from single particles has been observed, particularly for SERRS. However, the greatest enhancement occurs from interactions between particles. When silver nanoparticles are adsorbed onto a surface to form a layer, it is possible to study the distribution of enhancement across the surface. With any appropriate laser frequency, some parts of the surface become extremely active and other parts remain inactive. These active parts are called 'hot spots' and the regions which are active depend upon the excitation frequency used. It can be shown that the particularly active sites are at points between particles. The precise reasons for this are still the subject of research but the basic reasons are contained in the simple theory already described. Each individual particle will have a plasmon for which the resonance condition is only satisfied by a small range of wavelengths. However, the electrons are only loosely held and are free to couple to adjacent particles so that the plasmon is actually the plasmon of more than one particle and will have a new frequency range over which resonance can occur. The actual frequency of the plasmon for single particles decreases as the particle size rises and similarly dimers, trimers, etc., of particles have plasmon resonances at lower frequencies. The point at which two particles touch will generate enormous electric fields so that points of contact will give particularly effective SERS. Other more localized features may also contribute large amounts of scattering to the total. Thus, the low frequency chosen, the particle size and shape, and the way the particles organise in clusters will also contribute to the SERS enchancement.

Figure 5.2 shows a TEM of a typical colloid used in these experiments. It can be seen that there is a variation in particle size and that there are a number of particle interactions caused by the way that this material has been dried onto

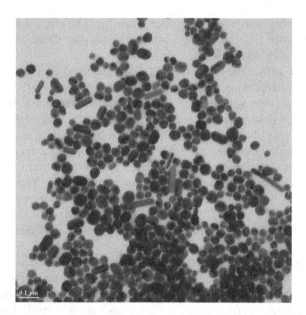

Figure 5.2. A typical colloid used in SERS. It was prepared by citrate reduction of silver nitrate and shows some needles which occur in some but not all preparations.

the surface. In some practical applications suspensions of colloidal particles are used so that scattering from many particles are detected and averaged.

5.3.2 Charge Transfer

Charge transfer or chemical enhancement [7] involves the formation of a bond between the analyte and the metal surface. This bond is believed to produce a surface species which includes the analyte and some surface metal atoms. This makes it possible to transfer charge (electrons or holes) from the metal surface into the analyte. The formation of this surface species will increase the molecular polarizability of the molecule considerably due to interaction with the metal electrons. Basically, the enhancement is thought to proceed via new electronic states which arise from the formation of the bond between the analyte and the metal surface. These new states are believed to be resonant intermediates in the Raman scattering. Thus, as opposed to the radiation being absorbed or scattered through the plasmons on the surface, the radiation is absorbed into the metal. A hole is transferred into the adsorbate metal atom cluster, the Raman process then occurs, excitation is transferred back into the metal and re-radiation occurs from the metal surface.

There is evidence for both these theories. However, it is very difficult to differentiate them. Clearly, chemical enhancement should occur only from molecules directly attached to the surface and consequently should increase only up to monolayer coverage. However, electromagnetic enhancement, although a longer range effect, drops off as $1/r^3$ with distance from the surface. Thus, most of the electromagnetic enhancement will also arise from adsorbates present on the surface up to monolayer coverage. The vast majority of evidence points to both effects having a part to play although it is generally believed that electromagnetic enhancement may have a greater part to play than charge transfer enhancement.

5.4 SELECTION RULES

SERS spectra are not straightforward to interpret. New peaks which do not appear in normal Raman scattering can appear in SERS and some peaks which are strong in normal Raman scattering can become very weak or disappear altogether. In addition, the intensity changes which occur at different concentrations can be nonlinear. A classic example of this is with pyridine. Well below monolayer coverage, the pyridine spectrum is very weak but it becomes quite strong as monolayer coverage is approached. At low concentrations the molecules are present on the metal surface with the plane of the pyridine ring parallel to the plane of the surface. As the concentration increases the plane of the pyridine ring is forced into an orientation perpendicular to the surface to allow more molecules to pack. This causes a rapid rise in SERS intensity. The reason this alters the intensity of SERS has already been discussed in general when it was stated that one requirement for scattering is that there is a polarizability component perpendicular to the surface. When light interacts with the surface, the effect can be described by two electric dipole components parallel and perpendicular to the surface. It is molecular polarizability caused by the perpendicular component which leads to scattering from the rough surface. For pyridine, the plane of the ring will produce the greatest polarizability changes. Thus, if the molecule is lying with the plane parallel to the surface, most of the polarizability change will be parallel to the surface and consequently will not contribute to scattering. When the plane is perpendicular to the surface the scattering process will be efficient.

The appearance of new bands further complicates the assignment. The most common reason this occurs is when a molecule has a centre of symmetry. Adsorption of the molecule onto a metal surface will effectively break the centre of symmetry. This results in the mutual exclusion rule (see Chapter 1) no longer being applicable, allowing some of the infrared active bands to break through and appear in the SERS spectrum. However, the situation is more complex than that. Some types of bands are naturally more intense in SERS than they are in normal Raman scattering. An example of this will be shown in Chapter 6. A consideration of the main effects led Creighton [6] to propose selection rules which work well in most circumstances. In theory they refer to electromagnetic enhancement but seem to be applicable in many cases. The problem with chemical enhancement is that the nature of the species formed between the adsorbate and the surface is not clearly defined. In principle, the selection rules should refer to the surface species including the molecule and the metal atoms complexed to it. However, in practice, considering the molecule as a distinct entity and ignoring the effect of the metal atoms other than the change they cause in molecular symmetry appears to be effective in most cases. These simple selection rules have proved to be useful in determining the orientation of a molecule on the surface and explaining some of the differences between SERS spectra and normal Raman spectra. However, there is still much to be learned about the reasons for the intensities of SERS active bands.

The fact that there are selection rules is a problem in SERS. The appearance of new bands and the disappearance of existing ones makes it much more difficult to relate the spectra obtained on the surface to that obtained from normal Raman scattering. In addition, since the sensitivity of SERS compared to normal Raman scattering is huge (a factor of 10^6), the dominant features of the spectrum could arise from a contaminant which sticks strongly to the surface. These problems make positive assignments difficult. Overcoming this difficulty is one of the key advantages of SERRS. However, before going on to explain the advantages and disadvantages of SERRS, a few applications of SERS are discussed.

5.5 APPLICATIONS OF SERS

SERS is one of the very few methods which can give effective, molecularly specific information about an adsorbate on a metal surface, *in situ*, in aqueous solutions. It has been applied to the detection of many molecules. For example, benzotriazole is widely used as an anti-corrosion agent for copper and an anti-tarnish agent for silver. Therefore the detection of the formation of the surface layer and elucidation of the nature of the surface complex are of considerable commercial importance. Roughened copper or silver surfaces treated with benzotriazole give very good SERS spectra which can be assigned positively to benzotriazole. The approximate orientation of the molecule and something about the nature of the complex it forms can be deduced. Of course, effective SERS is obtained only from certain surfaces of some metals and, as referred to earlier, even for some effective metals such as copper, it is difficult to study the surface reactions because of the rapidly changing nature of the copper surface in the presence of oxygen. One way of extending SERS to other surfaces is to add silver colloid to the surface of another metal which had been treated previously to form a surface layer of an organic molecule. This causes contact between the organic layer and the silver and does give effective spectra. However, it is difficult to assign the spectra and relate it to the structure of the organic surface layer, as it may not be clear whether the organic adsorbate remains attached to the original SERS-inactive surface or transfers to the silver surface. Thus, although the technique is simple, the interpretation requires careful consideration.

It is also possible to detect SERS from a non-metallic material. SERS from the surface of the polymer polyethylene terephthalate (PET) is of interest because it is often surface-treated with other polymer layers. Good SERS are obtained when a thin film of silver is deposited on the surface and the silver film irradiated from the side away from the polymer so that the radiation strikes the silver first. The spectra (Figure 5.3) are different from the Raman spectra, indicating the effect of surface selection rules on the spectrum. In particular, the band above $1700 \, cm^{-1}$, which is due to the carbonyl group in the Raman spectrum, is weaker in the SERS spectrum and in addition new bands appear. However one of the most notable features of this spectrum is the fact that it is taken by irradiating and collecting from the silver side away from the polymer. Since silver is a good absorber of visible light, no appreciable exciting radiation should penetrate the film into the polymer. Further, no weak Raman scattered light formed as a result of any breakthrough should be transmitted back through the silver. This appears to be the case in practice. Bands strong in

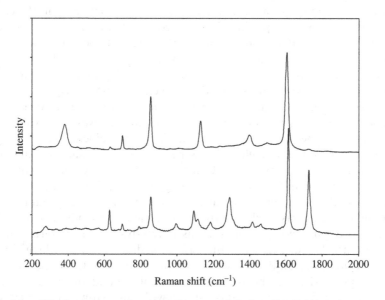

Figure 5.3. SERS spectrum of polyethylene terephthalate (PET) taken from a sample with a silver film cold-deposited on the surface (top) compared to a Raman spectrum of the same sample (bottom). SERS was recorded using excitation was from the side of the silver film away from the polymer layer.

normal Raman scattering from the polymer and weak in SERS do not appear except at the intensity expected for SERS. The reason this works is that the effective silver film is thin (approximately 15 nm) and the plasmon is much larger than this so that it transcends the film and is operative on both sides. Thus an interaction between the molecule and the plasmon on one side of the film can be detected on the other side. This results in only the surface layer in contact with the silver being enhanced and not the bulk of the polymer. Obtaining spectra from a very thin surface layer is difficult using simple techniques so that this is valuable information. In addition this example illustrates two key features of SERS which were discussed in the theory section. There are different selection rules compared to Raman scattering and there is definite coupling of the SERS-active species to the plasmon.

Another possible use for SERS is as a detection system in chromatography. An electrode can easily be placed in the line and can be cleaned by recycling and recreating the roughened surface after every determination. Provided the analyte or a suitable derivative adsorbs strongly on the metal surface, sensitive detection is possible. Alternatively, a flowing stream of colloid can be mixed with the effluent from the chromatography column and the stream passed through a capillary. The signal is then detected by irradiation and collection of the scattered light from the side of the capillary. However, there are major problems with SERS detection. With electrochemistry, it can be difficult to replicate the roughness on the electrode surface each time and, in the case of colloid, batch to batch reproducibility and control of the aggregation state are difficult. The colloidal technique is proving in practice to be more sensitive, and considerable efforts are being devoted to obtaining greater reliability.

With careful control, colloid can be made repeatably to a particular specification and TEMs indicate this colloid can be close to mono-disperse. Provided the colloid is time stable, a standard can be used to calibrate out variations in the assay and so a degree of quantitation is possible. However, only molecules which adhere to the surface will be active. The main problem is controlling the state of aggregation of the colloid. In essence, the colloid is stable because of the surface charge on the particles. Commonly, the aggregating agent added reduces this charge and causes controlled aggregation. This shifts the plasmon frequency so that some aggregates are in resonance with the laser but remain in suspension. The agents used vary from inorganic compounds such as sodium chloride to organic compounds such as poly-L-lysine. They work either by reacting with the surface, as in the case of sodium chloride which forms a silver chloride layer on the surface, or simply by coating the surface as is the case with poly-L-lysine. This produces less stable colloid and causes aggregation. It is a dynamic process that continues with time. However, in practice, it is possible to produce suspensions which are sufficiently stable to give time-stable SERS for periods of up to several hours in some cases. Given that the SERS measurement is very sensitive, and accumulation times of 10 s are normal, this should be adequate. However, it is important that this aggregation process is controlled as effectively as possible. One way this has been done is by using a flowcell. In this system the colloid flows down one tube and the aggregating agent down another. Once thoroughly mixed the analyte is added down a third tube to the aggregated colloid and the flowing stream is interrogated by the Raman spectrometer. In this system, relative standard deviations of less than 2% can be obtained and the system is sufficiently quantitative for regular use.

Given that we know that the enhancement from each molecule may well be different, the reason this technique can give quantitative results is because it averages over many molecules and many hot spots. A very different effect is found when very small numbers of molecules are present. At this point, the Raman signal does not remain constant during accumulation of the scattering. Every time a hot spot with an adsorbate on it passes through the beam, a burst of scattered light is recorded. This process has some similarities to true single molecule detection where signal 'blinking' is observed but it is more about single aggregate detection. Thus, claims of single molecule detection by this technique alone should be treated with some reserve unless well supported by other evidence.

Because of the averaging effect, a good analyst can obtain quantitative information from SERS. Figure 5.4 shows the results of the addition of amphetamine to a colloidal suspension aggregated with sodium chloride. As

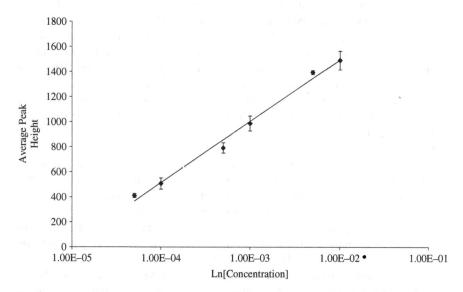

Figure 5.4. SERS from amphetamine added to a colloidal silver suspension aggregated with sodium chloride. (Reproduced with permission from K. Faulds, W.E. Smith, D. Graham and R.J. Lacey, *Analyst*, **127**, 282 (2002).)

can be seen, the results are quantitative over two orders of magnitude and provide a sensitive method of detecting amphetamine in solution.

Thus, if SERS is to be used as a detection technique, a careful consideration of the chemistry of the surface and the physics of surface enhancement is required; but with care and by using the technique within its limitations, good quantitative measurements can be obtained. Overall, the difficulties with SERS have certainly limited its application. However, the related technique of SERRS has fewer problems and such unique properties that it is likely to become an important analytical method for the solution of specific problems.

5.6 APPLICATIONS OF SERRS

As stated above, SERRS combines the advantage of surface enhancement with the use of a resonant chromophore so that molecular resonance enhancement as well as surface enhancement is obtained. In this method, a dye or a molecule containing a dye is used as the analyte. The enhancement obtained is probably due to one single process and cannot be explained simply by adding molecular resonance and surface enhancement together. For example enhancements of up to 10^{14} have been claimed rather than 10^8 or 10^9 and users certainly experience much greater sensitivity with it than would be expected simply from the addition of molecular resonance to SERS. Having said this, it is easier experimentally to treat the effect as being due to a combination of the two known effects; provided the previous caveat is borne in mind, this works well. There are a number of reasons why the technique is effective. The absorption of the dye directly onto a metal surface causes all fluorescence to be quenched. This is a very efficient mechanism, more efficient than that of standard molecular quenchers, and consequently both fluorescent and non-fluorescent dyes can be used as analytes or as part of analytes.

The signals obtained from SERRS often resemble quite closely those obtained from resonance Raman scattering with much less evidence of orientation dependence or other surface selection rules. Normally, the relative intensities of bands in the SERRS spectrum and in the resonance spectrum may differ by up to 30% but any frequency shifts are very small. Thus, normally it is easy to recognize a particular adsorbate in SERRS simply by comparing it with the resonance Raman spectrum or, in the case of a fluorescent compound, the FT Raman spectrum with the SERRS spectrum. Further, there is less of a problem with contaminants since the additional resonant enhancement discriminates against many of them. Finally, the additional enhancement often means that lower laser powers and shorter accumulation times can be used. Consequently surface photodecomposition is much less of a problem.

The increased reliability with which a species can be positively identified provides increased confidence when measuring at very low concentrations that the signals obtained do come from the analyte. This in no way means that contamination is no longer a problem with the technique. All extremely sensitive analytical methods suffer from the possibility that other molecules may be present in extremely small quantities; for example, adsorbed on a vessel wall from a previous experiment, from rubber gloves, or from seals or stoppers. However, the fact that they can be instantly recognized is a significant advantage over most other ultrasensitive techniques, and leads to confidence in the reliability of SERRS.

5.7 THE BASIC METHOD

Ideally, the analyte should be a dye which has an absorption maximum at or close to the frequency of the plasmon resonance on the active surface. In this way both molecular resonance and surface plasmon resonance are obtained. In practice this is not always possible or, in some cases, even desirable. However, it appears in practice that an exact match is not required. Figure 5.5 shows a simple diagram explaining the different ways in which SERRS can be obtained. In Figure 5.5a the plasmon band and the absorption band are at the same frequency and so the excitation should be chosen to be close to that frequency. In Figure 5.5b the absorption band and the plasmon band are at different frequencies and the choice would have to be made as to whether to choose the excitation frequency at the frequency of the plasmon resonance, at the frequency of the molecular resonance or in between.

Using colloidal silver as substrate, SERRS has been recorded from a nonaggregating dye with no aggregating agent. The SERRS must come from single particles and consequently does not include any enhancement from hot spots or particle/particle interactions. Nevertheless, in contrast to SERS the signals are quite strong. The strongest signals were obtained at the plasmon resonance maximum, with much smaller signals obtained at the dye absorption maximum. It would appear that for single particle excitation, the frequency chosen would be that of the surface plasmon.

With an aggregating dye, the absorption maximum in the electronic spectrum at the plasmon frequency of the single particle is reduced and a broad absorption at longer wavelengths grows in. This is due to the formation of a range of species including dimers, trimers and a range of clusters of different sizes. Each of these species has a separate absorption maximum and the broad band is a combination of absorbances from each of them. At any one of these frequencies, some but not all of the aggregated species will have a plasmon which is in resonance and so these species contribute significantly to the enhancement. In this case, the combination of surface enhancement and

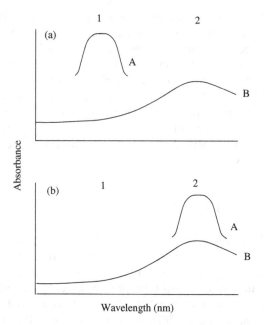

Figure 5.5. Different combinations of plasmon resonance (B) and absorption band (A) arrangements for SERRS. (Reproduced with permission from C. Rodgers, W.E. Smith, G. Dent and M. Edmondson, *J. Chem. Soc. Dalton Trans.*, 791 (1996).)

resonance enhancement extends over a range of frequencies. In practice this is a range up to 150 nm or so from the molecular resonance frequency above which the enhancement tails off to the level expected for SERS.

Figure 5.6 shows a plot of the intensity of a major peak from two similar azo dyes, one which does not cause aggregation (dye A) and one which does (dye B). The absorption maxima of the dyes were 429 and 442 nm, respectively, and the absorption maximum of the colloid unaggregated was 404 nm. Dye A contains a hydroxyl group which ionizes at neutral pH to give a negative ion. When adsorbed on the surface this helps maintain the negative charge of the particle and prevents aggregation. Dye B displaces negatively charged material on the surface and reduces the surface charge. With dye A the SERRS intensity was greatest at the frequency of the surface plasmon and dropped off away from it, as expected for unaggregated colloid. Dye B showed greater intensity across the visible region dropping off towards the infrared, as expected for aggregated colloid.

The use of coloured reagents to obtain the extra sensitivity has led to a large number of single molecule experiments. It is reasonably clear now that SERRS has the potential to detect a single molecule and that the molecule can be positively identified in solution. Consequently SERRS has some unique advantages

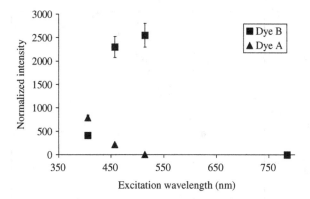

Figure 5.6. Plot of the intensity dependence of one peak for both a non-aggregating dye (single particle enhancement) and an aggregating dye (cluster enhancement). (Reproduced with permission from K. Faulds, R. Littleford, D. Graham, G. Dent and W.E. Smith, *Anal. Chem.*, **76**, 5902 (2004).)

for ultra-sensitive analysis. However, it has also been shown that as with SERS, there are hot spots and the greatest enhancement is obtained at points where two nanoparticles touch. In addition, there are several studies which show that, as for SERS, the enhancement appears to be uneven across the surface. Thus, although some of the problems with SERS have been overcome, care must still be taken if reliable representative results or quantitative results are to be obtained. Two examples which give quantitative analysis and illustrate the practical advantages are given below.

The concentration of the coloured anti-cancer drug mitoxantrone is difficult to analyse in blood. It requires quite a complex procedure usually involving separation of the mitoxantrone from the blood serum and chromatography. Using a flow system to control aggregation, a drop of plasma from a patient who had been undergoing a course of therapy with mitoxantrone was added to the flow cell. Not all the plasma would be expected to adsorb on the silver colloid and consequently a fluorescence background might be expected from non-adsorbed material. However, when the flowing stream containing the plasma was mixed with streams containing colloid and an aggregation agent and the mixed material passed across the spectrometer, excellent mitoxantrone spectra were obtained with little evidence of background fluorescence. One reason is that the flow cell causes a significant dilution in the plasma before measurement and this dilutes out the background fluorescence from the non-adsorbed material. However, the mitoxantrone desorbs from the plasma proteins and adheres strongly to the silver surface. Thus, SERRS from mitoxantrone can be obtained within a minute or less directly from serum sample. Figure 5.7 shows results over a range of concentrations. There is good linearity

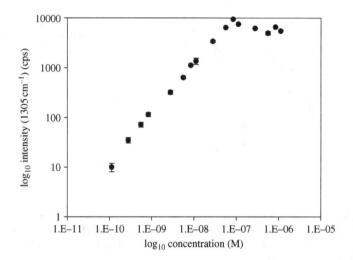

Figure 5.7. Intensity dependence of one peak from the SERRS spectrum of mitoxantrone in plasma. (Reproduced with permission from C. McLaughlin, D. MacMillan, C. McCardle and W.E. Smith, *Anal. chem.*, **74**, 3160–3167 (2002).)

over orders of magnitude. This range encompasses the effective concentration range found in patients so that the technique can be used directly. This is an example where SERRS could be considered as the technique of choice. However, this assay used a coloured drug which adsorbed strongly onto the silver surface. Doing this with other drugs could be more difficult.

One big advantage of SERRS is that it gives a set of sharp, molecularly specific peaks and therefore it should be relatively simple to differentiate several SERRS-active analytes without separation. For example, this would mean that it would be possible to write codes into a piece of plastic or to detect several DNA strips without separation. Using a simple Raman spectrometer with a telescopic lens and a laser providing 3.6 mW of 532 nm excitation, SERRS was detected at distances of up to 20 m within 10 s. The sample was colloid-treated with a dye and fixed in a polymer. This result is illustrated in Figure 5.8. It indicates that the SERRS sensitivity can be obtained in practice.

With DNA, effective SERS is difficult to obtain since the DNA molecule is negative and most of the colloidal particles used for this type of analysis are also negative. Thus, spectra can be obtained at higher concentrations but not at really low concentrations. In addition DNA replicates the same four bases and a sugar and often it is the characterization of the order of these bases or some defect in one of the many which is required. Consequently sharp information on any one base can be difficult to obtain. The same problem is encountered in other spectroscopies. To overcome this problem, one standard approach is to use sequences of DNA which will recognize the complimentary strand that

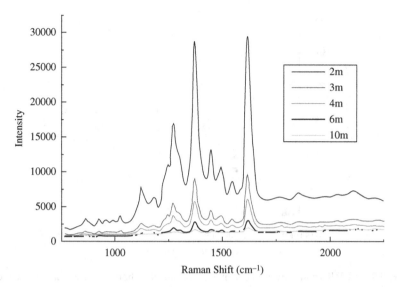

Figure 5.8. Distance dependence of the SERRS spectrum of a dye adsorbed onto silver colloid incorporated in a polymer. (Reproduced with permission from Ailie McCabe, W.E. Smith, G. Thomson, D. Batchelder, R. Lacey, G. Ashcroft and B.G. Foulger, *Appl. Spectrosc.*, **7**, 56 (2002).)

contains the defect. These sequences are labelled with a fluorophore or any other marker. It is the marker which is detected in the analytical procedure. There are advantages in sensitivity and selectivity in using SERRS as the detection technique for these markers. To do this, DNA was modified by the addition of dyes that adsorb strongly onto the silver surface. Initially this was easier to do using fluorophores since they are commercially available coupled to DNA bases for use in fluorescence analysis. Using similar procedures with a standard fluorescence assay, DNA concentrations down to between 10^{-12} and 10^{-13} M could be determined; the results are shown in Figure 5.9 for one sequence. These results were obtained from a suspension of DNA in a cuvette. However, the Raman beam is sharply focussed and consequently only a very small volume of that cuvette is actually used for the analysis. If a calculation is done on how many molecules are present in the beam at any one time, the results indicate that we are at the single molecule sensitivity limit obtained by using a standard commercial spectrometer, normal laboratory methods and an accumulation time of 10 s. Good quantitative SERRS can now be obtained in 100 ms and the limitation here is not the SERRS technique, but the limitations of the instruments used to give quantitative results. Much shorter times can easily be achieved from the huge signals that are now routinely and reliably obtained.

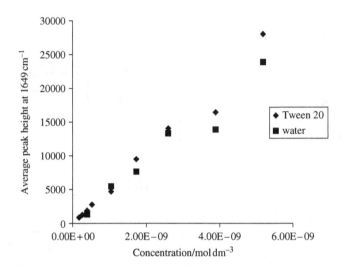

Figure 5.9. SERRS concentration dependence of one of eight dye-labelled oligonucleotides.

In one recent experiment, using DNA, eight different fluorophores were attached to DNA strips. From all but one of these SERRS, exhibited a sensitivity three to four orders of magnitude greater than that achieved by fluorescence detection using two standard fluorescence spectrometers. This is a very remarkable result. The SERRS experiment gives the full spectrum whereas the fluorescence method accumulates all the fluorescence through a simple and efficient filter system. It indicates clearly that SERRS has very significant potential for ultra-sensitive detection. It may still be argued by some that SERRS can be variable. In this particular experiment, the data quoted are detection limits which means that we have not only linear gradients but also relative standard deviations at the lowest levels which are better than those achieved through fluorescence, indicating the technique is now reliable and quantitative.

Another remarkable feature of this work is that it was done in a cuvette. Techniques which control aggregation such as flow cells could be used to improve the situation further.

The sharp, molecularly specific nature of the signals is another characteristic of SERRS. It can be used to label colloidal particles so that they are uniquely coded with the mixture. Figure 5.10 shows the result of mixing three different dyes and a dye labelled oligonucleotide together into a suspension of silver colloid. Each dye can be separately identified in the suspension. Of course there is no guarantee that all dyes are on any one particle, but the experiment shows that with further development, SERRS has great potential for coding.

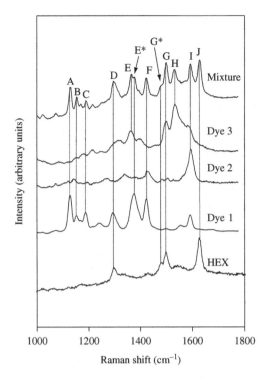

Figure 5.10. A colloidal suspension containing three dyes and an oligonucleotide illustrating the ability of SERRS to discriminate between labels in solution.

There are no particular secrets in this work. The technology is now relatively simple and readily available. What is important is that both the chemistry of the attachment to the surface and the physics of the measurement must be considered by the analyst.

REFERENCES

1. M. Fleischman, P.J. Hendra and A.J. McQuillan, *Chem. Phys. Lett.*, **26**, 163 (1974).
2. D.C. Jeanmarie and R.P. Van Duyne, *J. Electroanal. Chem.*, **84**, 1 (1977).
3. M.G. Albrecht and J.A. Creighton, *J. Am. Chem. Soc.*, **99**, 5215 (1977).
4. M. Moskovits, *Rev. Mod. Phys.*, **57**, 783–826 (1985).
5. A. Campion and P. Kambhampati, *Chem. Soc. Rev.*, **27**, 241 (1988).
6. J.A. Creighton, in: *Spectroscopy of Surfaces*, R.J.H. Clark and R.E. Hester (eds), Wiley, 1998, p. 27.
7. A. Otto, I. Mrozek, H. Grabhorn and W. Akemann, *J. Phys. Cond. Matter*, **4**, 1142 (1992).

Chapter 6

Applications

6.1 INTRODUCTION

In the previous chapters, examples of materials examined by Raman spectroscopy have been given to highlight specific aspects of the technique. Whilst the technique was initially used to examine inorganics, it grew with extensive use in polymer analysis. More recently there has been a growth in pharmaceutical applications, while other applications have been successfully established in colours, semiconductors, art, archaeology and biotech areas. There have also been advances in forensic and process analysis. In several areas the ability of Raman spectroscopy to analyse materials in glass, water, inside packaging materials or *in situ* directly has been exploited. This chapter will attempt to show ways in which Raman spectroscopy can be used to solve specific analytical problems. Compared to many other analytical techniques Raman spectroscopy is viewed by many as a niche technique. However the applications are fast growing and in this chapter we can only attempt to highlight the areas where the technique has been successful. Some applications within the authors' experiences are given in more detail to exemplify the strengths and pitfalls of the technique which will act as a guide to the reader's potential use in their own areas of interest.

6.2 INORGANICS AND MINERALS

Raman spectroscopy is a very good tool for the examination of inorganic materials or those containing inorganic components. Raman spectroscopy is the one of the few analytical techniques which can positively identify and characterize both elements and molecules. Raman spectroscopy can unambiguously identify

Modern Raman Spectroscopy – A Practical Approach   W.E. Smith and G. Dent
© 2005 John Wiley & Sons, Ltd   ISBNs: 0-471-49668-5 (HB); 0-471-49794-0 (PB)

both the purity and physical form of elemental carbon, germanium, sulphur and silicon, the last mentioned being very important in the semiconductor industry as described in Sections 6.6 and 6.9. Raman spectroscopy is the only analytical technique which can positively identify and characterize elemental carbon from the shape and position of the bands in the spectrum. Starting with amorphous carbon, the bands sharpen as the crystallinity increases and then reach the ultimate with pure diamond giving a single sharp band at $1365 \, cm^{-1}$. However the ability to carry out this examination is dependent on the exciting wavelength. The spectrum of diamond can easily be recorded at 1064 nm excitation (Figure 6.1) but amorphous carbon usually causes strong absorption and burning.

Spectra of amorphous carbon and graphite have been readily recorded with visible excitation [1, 2], but they can be recorded at 1064 nm only by employing dilution techniques. Elemental sulphur is another strong Raman absorber with strong bands at $\sim$200 cm^{-1}. It is sometimes used as a standard for instrument performance checks. The spectra do however vary with physical form. Flowers of sulphur (monoclinic) have a different spectrum from other sulphur forms [3]. Figure 6.2 shows both Stokes and anti-Stokes bands of sulphur.

Early work carried out in Raman spectroscopy showed its strength with inorganics. Particulates in urban dust [4] such as anhydrite, calcite, dolomite and quartz have been identified and characterized. Early microprobes [5, 6] were used to record the spectra of gaseous, liquid and solid inclusions in minerals. These include $CH_4/CO_2/N_2$ ratios and solids such as apatite, calcite, nacholite and sulphur [7]. Inorganics have also been identified in biological samples, e.g. copper sulphide needles inside the lysosomes of *Littorina littorea* [8].

Figure 6.3 quite clearly shows the differences in the Raman spectra of the rutile and anatase forms of titanium dioxide, an important industrial filler for

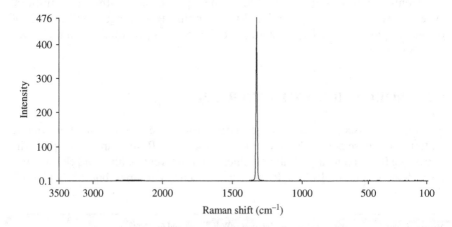

Figure 6.1.   NIR FT Raman spectrum of diamond.

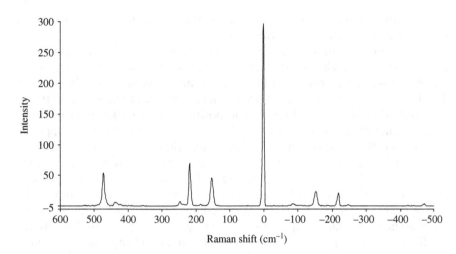

Figure 6.2. NIR FT Raman spectrum of sulphur showing both Stokes and anti-Stokes shifts.

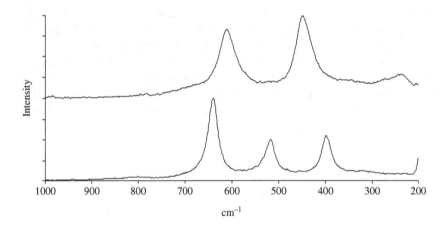

Figure 6.3. NIR FT Raman spectra of TiO_2 – rutile (top); anatase (bottom).

polymers. This difference has been employed quantitatively for plant control as described in Section 6.9. The breadth of the TiO_2 bands is quite unusual for the Raman spectrum of an inorganic compound. The vast majority of inorganic compounds have spectra with very sharp bands. The sharp bands make them relatively easy to pick out in the spectra of other compounds. In many cases the Raman spectra of organic materials have strong bands in the same position as in the infrared spectrum; with inorganic materials there are some notable exceptions. In the spectra of sulphates the bands have very different shapes but are in similar positions in the spectrum, whilst the carbonate bands are significantly moved (Figure 6.4). This is due to the relative intensities of symmetrical and asymmetrical vibrations.

A list of band positions observed in some common inorganic compounds is given in Table 6.1. The spectra were recorded with a 1064 nm exciting wavelength. Copies of the spectra, in PDF format, are available on the Internet [9]. This is one of the few application areas where there are published texts [10, 11] on a collection of specific Raman band positions. Using other exciting wavelengths generally does not affect the band positions but there are apparent exceptions [12]. Besides the normal background fluorescence, specific bands in minerals have been reported due to fluorescence. These can be misinterpreted as chemical group bands. Features to be aware of in studying the Raman spectra of inorganic materials are the changes which can occur due to the form or orientation of the crystal in the beam. Inorganic compounds tend to be more crystalline than many organic compounds and hence more susceptible to these effects. As discussed in Chapter 2 particle size effects can also change the spectrum.

Minerals are naturally occurring inorganic molecules with specific structure and form. As stated previously, a lot of early Raman spectroscopy and microprobe work was carried out on minerals, the latter for the identification of impurities and inclusions [13]. This area has found geological applications in studying both terrestrial [14] and extra-terrestrial [15] materials. Again, tables of known band positions have been published for minerals [16].

There has been a slow but growing increase in the use of Raman spectroscopy for commercial applications of both inorganic and mineral analysis: TiO_2 plant monitoring, which has already been mentioned, quantitative methods for inorganics in storage tanks [17], diamond quality checks [18, 19] and testing of various jade minerals [20].

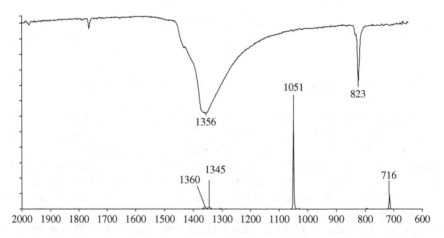

Figure 6.4.   Infrared and NIR FT Raman spectra of $NaCO_3$.

Table 6.1. List of band positions in cm^{-1} observed in some common inorganic compounds. Bold type indicates the strongest bands

Ammonium	Carbamate	**1039**							
Diamond	Carbon	**1331**							
Ammonium	Carbonate	**1044**							
Calcium	Carbonate	**1087**	713	282					
Lead (II)	Carbonate	1479	1365	**1055**					
Potassium	Carbonate	**3098**	1062	1062					
Strontium	Carbonate	1072							
Potassium	Carbonate (99.995%)	**1062**	687						
Potassium	Carbonate (99.995%), rotator	**1061**	686						
Sodium	Carbonate (anhydrous)	**1069**							
Sodium	Carbonate (anhydrous), rotator	**1080**	701						
Sodium	Carbonate AR	1607	**1080**	1062					
Sodium	Carbonate monohydrate	**1070**							
Potassium	Carbonate (99.995%), rotator	**1061**							
Potassium	Carbonate, Aldrich, 99%	**3098**	1062	1062					
Sodium	Chloramine-T, sodium salt	3069	2921	1600	1379	1213	**1132**	930	800
Sodium	Dichloroisocyanurate	1733	1051	707	577	**365**	230		
Potassium	Dichromate	**909**	571	387	235				
Sodium	Dichromate (2H$_2$O)	**908**	371	236					
Potassium	Dichromate, rotator	**909**	570	374	235				
Potassium	Dichromate, rotator	**909**	570	374	235				
Ammonium	Dihydrogen orthophosphate	**925**							
Ammonium	Dihydrogen orthophosphate	**923**							
Potassium	Dihydrogen orthophosphate	**915**							
Titanium	Dioxide (anatase)	**639**	516	398					
Titanium	Dioxide (rutile)	610	**448**	237					
Sodium	Dithionite	1033	364	**258**					
Sodium	Dithionite	1033	364	**258**					
Ammonium	Ferrous sulphate (6H$_2$O), rotator	**982**	613	453					
Sodium	Hexametaphosphate	**1162**							
Ammonium	Hydrogen carbonate	**1045**							

Table 6.1. Continued

Caesium	Hydrogen carbonate	**1012**	671	634			
Potassium	Hydrogen carbonate	1281	**1030**	677	636	193	
Sodium	Hydrogen carbonate	1269	**1046**	686			
Di-ammonium	Hydrogen orthophosphate	**948**					
Di-ammonium	Hydrogen orthophosphate	**948**					
Di-sodium	Hydrogen orthophosphate	1131	1065	**934**	560		
Di-potassium	Hydrogen orthophosphate (trihydrate)	1048	**950**	879	556		
Potassium	Hydrogen sulphate	1101	**1027**	855	581	412	327
Sodium	Hydrogen sulphate	**1065**	1004	868	601		
Sodium	Hydrogen sulphate (monohydrate)	1039	857	603	412		
Calcium	Hydroxide	1086	**358**				
Sodium	Hydroxide	**205**					
Lithium	Hydroxide (monohydrate)	1090	839	517	397	**213**	
Ammonium	Hydroxy chloride	1495	**1001**				
Potassium	Iodate	**754**	660				
Sodium	Metabisulphite	**1064**	660	433	275		
Barium	Nitrate	1048	733				
Bismuth	Nitrate	**1037**					
Lanthanum	Nitrate	**1046**	739				
Lithium	Nitrate	1384	**1070**	735	237		
Potassium	Nitrate	1051	716				
Silver	Nitrate	**1046**					
Sodium	Nitrate	1386	**1068**	725	193		
Magnesium	Nitrate (6H$_2$O)	**1060**					
Iron(III)	Nitrate (9H$_2$O)	**1046**					
Potassium	Nitrite	1322	**806**				
Silver	Nitrite	**1045**					
Sodium	Nitrite	**1327**	828				
Sodium	Nitrite	**1327**	828				
Silver	Nitrite, rotator	**1045**	847				
Sodium	Nitroprusside (2H$_2$O)	2174	1946	1068	656	**471**	

Compound	Form								
Tri-potassium	Orthophosphate	**1062**	940						
Tri-sodium	Orthophosphate	941	415						
Tri-sodium	Orthophosphate	1005	**940**	548	417				
Tri-potassium	Orthophosphate	**1062**	972	857	549				
Tri-sodium	Orthophosphate (12H₂O)	939	407						
Tri-sodium	Orthophosphate (12H₂O)	940	550	413					
Tri-potassium	Orthophosphate (H₂O)	**1061**	939						
Tri-potassium	Orthophosphate (H₂O), rotator	**1061**	940						
Cupric	Oxide	296							
Zinc	Oxide	438							
Cupric	Oxide, rotator	297							
Zinc	Oxide, rotator	439							
Magnesium	Perchlorate	964	643	456					
Ammonium	Persulphate	**1072**	805						
Potassium	Persulphate	1292	**1082**	814					
Sodium	Persulphate	1294	**1089**	853					
Sodium	Phosphate	938							
Calcium	Silicate	983	578	373					
Lithium	Silicate	601							
Zirconium	Silicate	3019	**2821**	2662	1004	438	355	197	
Lithium	Silicate	**589**							
Calcium	Silicate hydrous, commercial	983	578	372					
	Rotator	677	**195**						
Magnesium	Silicate hydrous (talc)	676	**194**						
Magnesium	Silicate hydrous (talc), rotator	677	362	195					
Aluminium	Silicate hydroxide (kaolin)	**466**							
Aluminium	Silicate hydroxide (kaolin), rotator	912	791	752	705	**473**	430	338	276
Ammonium	Sulphate	**975**							
Barium	Sulphate	**988**	454						
Barium	Sulphate	**988**	462						
Calcium	Sulphate	1129	**1017**	676	628	609	500		
Magnesium	Sulphate	**984**							
Potassium	Sulphate	1146	**984**	618	453				

Table 6.1. Continued

Silver	Sulphate	**969**							
Sodium	Sulphate (anhydrous)	**993**							
Calcium	Sulphate (dihydrate)	1135	**1009**	669	629	491	415		
Zinc	Sulphate (heptahydrate)	**985**							
Barium	Sulphate, Raman microscope	**986**	458						
Barium	Sulphate, rotator	**988**	462						
Barium	Sulphate, static	**988**	462						
Sodium	Sulphite	**987**	950	639	497				
Potassium	Sulphite	**988**	627	482					
Magnesium	Thiosulphate (hexahydrate)	1165	1000	659	**439**				
Sulphur	–	471	**216**	151					
Barium	Thiosulphate	1004	687	**466**	354				
Potassium	Thiosulphate (hydrate)	1164	1000	667	**446**	347			
Sodium	Thiosulphate (pentahydrate)	1018	**434**						
Potassium	Titanium oxalate	1751	1386	1252	850	**530**	425	352	300
Potassium	Titanium oxalate (2H$_2$O)	1751	1384	1253	851	**526**	417	353	299

6.3 ART AND ARCHAEOLOGY

One of the major advantages of Raman spectroscopy in this field comes from the difficulties of obtaining samples for analysis. The materials to be examined are either very valuable in themselves or part of an object which is extremely valuable. Removing even the smallest sample for analysis would cause damage and subsequent loss of value. Raman spectra can be obtained from microsamples or by use of confocal techniques from under layers without having to separate them. When even microsamples cannot be taken, then examination can take place using sensing heads at the end of fibre optic probes and/or by remote sensing [21]. There is extensive knowledge of the compositions of colour used in paintings and decorated *objects d'art*. For many centuries colouring came from inorganic pigments and natural dyes. There were very few synthetic dyes available until the 19th century. Raman spectroscopy can not only identify the type of inorganic materials used but also the physical forms. The use of the dyes, pigments and resins [22] can be established chronologically [23]. By examining the Raman spectra of paintings [24] and archaeological artefacts such as pottery [25] the age of the work can be determined. Figure 6.5 demonstrates this feature very well. The knowledge of the composition can be used to distinguish original work from restoration and/or forgery.

Besides colour identification, the identification of gemstones [26], porcelains [27], metal corrosion products [28] and organic materials such as resins [29] and ivory [30] makes Raman spectroscopy a very valuable technique in this field. Ivory, of particular interest, has been studied for environmental purposes and by law enforcement bodies. Vibrational spectroscopic assignments of mammalian ivories have been published [31].

6.4 POLYMERS AND EMULSIONS

6.4.1 Overview

The applications of Raman spectroscopy to polymers are extensively reported in the literature. A recent publication [32] devoted 10 chapters to the vibrational analysis of polymers with Raman spectroscopy being extensively used. Only a general overview is given here in an attempt to highlight some of the strengths. Polymers have been studied for identification, structure, composition, cure and degree of polymerization in the solid, melt, film and emulsion states. Ironically, polymers, particularly aliphatic ones, are not very strong Raman scatterers and sample preparation techniques such as folding thin films have to be employed. Conversely items such as aspirin can be studied, by Raman spectroscopy, inside a film wrap without interference from the film. In the early days Raman studies of polymers were restricted by fluorescence and thermal absorption due to

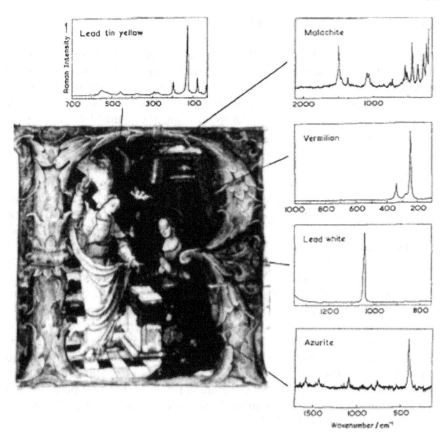

Figure 6.5. Sixteenth-century German choir book: historiated letter 'R'. (Reproduced with permission from R.J.H. Clark, *Journal of Molecular Structure*, **347**, 417–428 (1995).)

impurities and fillers. These have been largely overcome by NIR visible, UV dispersive excitation and 1064 nm FT excitation. Cleaner processes leading to fewer residues have also helped. However some commercial products still contain anti-oxidants, plasticizers and fillers which can cause interferences. Both the chemical and the physical nature of the polymers themselves have to be considered before examination by Raman spectroscopy. Whilst Raman spectroscopy can be considered a minimal sample preparation technique, the physical state (granules, film, etc.), the morphology (macro- and micro-crystallinity), thermal properties (high/low melting), state of cure, copolymer distribution and homogeneity of fillers can all affect the way in which a sample is presented to the Raman spectrometer. Many samples can be placed directly in the beam for 90° or 180° signal collection. Samples can be presented in glass bottles and aqueous emulsions can be studied. However, for the latter 'particle' size relative to the exciting wavelength needs to be considered. Care has to be taken if, in order to detect weak scatters the laser power is increased. This may cause thermal damage or induce changes.

6.4.2 Simple Qualitative Polymer Studies

Of the many in-depth and wide ranging reviews written on Raman spectroscopy of polymers, a few relatively simple applications are described here to illustrate the range of the technique.

Whilst collections of Raman spectra are not common, a published collection of spectra of common polymers is available [33, 34]. The spectra of five of the most commonly encountered polymers, polyethylene (PE), polypropylene (PP), polyethylene terephthalate (PET), polycarbonate and polystyrene were all recorded quickly and easily on an FT Raman spectrometer with no sample preparation. In Figure 6.6 the spectra are quite distinctive even to subtleties of chain branching between PE and PP in the 2900 and $1450 \, \text{cm}^{-1}$ region of the spectrum. Figure 6.7 shows that whilst the polycarbonate spectra are dominated by the bands due to the aromatic groups, differences between the aliphatic methyl and cyclohexyl groups are quite clear. The bands are in the same

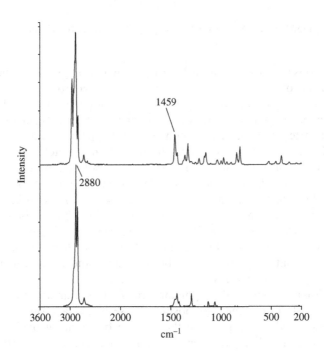

Figure 6.6. NIR FT Raman spectra of polypropylene (top); polyethylene (bottom).

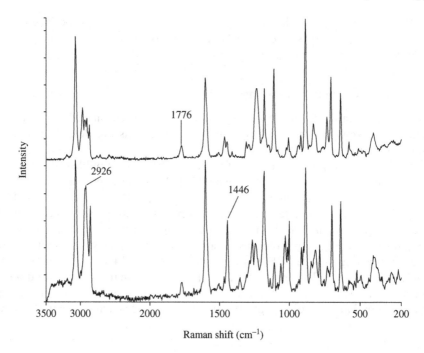

Figure 6.7. NIR FT Raman spectra of two polycarbonates.

spectral regions as in the PE and PP spectra shown in (Figure 6.6). The spectrum of PET, also shows a strong band due to the carbonyl group (Figure 6.8). This quite distinctive spectrum is also presented in Section 6.5 where the PET is seen as part of a matrix. These spectra also give the lie to the often perpetrated myth that asymmetric groups such as carbonyls do not appear in Raman spectra. The band at 1776 cm^{-1} in Figure 6.7 is due to the carbonyl stretch. In Figure 6.9 the comparison of infrared and Raman spectra of polystyrene does show how Raman spectra emphasizes the bands from aromatic groups. Because of the ability to show these sometimes subtle differences, Raman spectroscopy has been used to create microscopic images of the distribution of polymers in blends [35]. Differences in morphology, polymer chain ordering and molecular orientation have been the subject of in-depth studies [36–41]. One of the most effective combinations is to employ both infrared and Raman imaging tech-niques to study these features [42].

In addition to the study of the polymers themselves, Raman spectroscopy has been used to study polymer composites. These often have other components added for strength or to preserve the lifetime by reducing oxidation or free radical attack. The fillers are often inorganics such as silicates, carbonates and elemental carbon or sulphur. The Raman spectra of these, as seen earlier, are

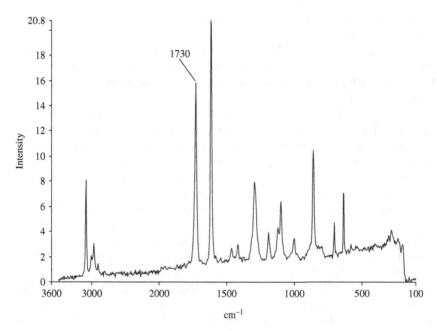

Figure 6.8. NIR FT Raman spectrum of PET.

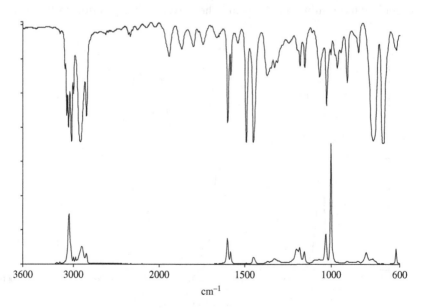

Figure 6.9. Spectra of polystyrene – infrared transmission spectrum (top); FT Raman (bottom).

quite distinctive. The variation in spectra can give information on the chemical and physical composition [43] and strength of the composite.

Raman spectra can identify the type of polymer present in a material but can also be used to study how far polymerization has occurred [44, 45] or even how far a polymer has been degraded [46, 47]. The double bond in acrylates is very strong and quite characteristic. As polymerization occurs the strength of this bond decreases. This can be easily monitored by Raman spectroscopy and has been developed for a number of applications for plant scale monitoring. Acrylate-based emulsions are quite common. In another example, the variation in the band due to $>C=C<$ at $1655\,cm^{-1}$ can be seen in Figure 6.10 for sunflower oil. The top spectrum is an emulsion with degraded sunflower oil. The bottom is a Raman spectrum of the pure oil.

Following the polymerization of dispersions of polymer in aqueous media is very difficult by conventional methods. However, by using a fibre optic probe, following the reduction of a band from the monomer by Raman scattering is quite feasible, in glass vessels under manufacturing conditions. Whilst this experiment is relatively easy to carry out in principle, one must consider the relative size of the emulsion drops in the suspension and the laser wavelength employed. This is similar to the particle size effects discussed in Chapter 2. The opposite of following a polymerization reaction is to monitor degradation of a polymer. One of the earliest reported degradation studies [48] was on polyvinyl chloride (PVC). As PVC degrades, hydrogen chloride (HCl) is lost and conjugated double bonds form. The wavenumber position in the Raman

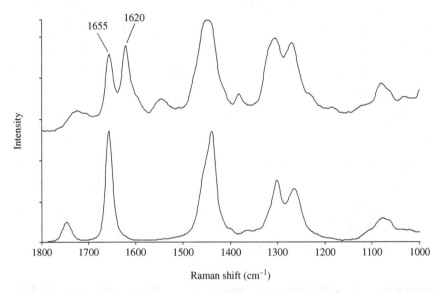

Figure 6.10.   NIR FT Raman spectra of oils – emulsion (top); pure oil (bottom).

spectrum shows the number of bonds in a chain which have been formed and hence the degree of conjugation and degradation. The studies by Gerrard and Maddams showed intense bands at 1511 and 1124 cm^{-1} associated with conjugation unsaturation. By variation of the illuminating laser line they showed that the intensity was due to resonance effects. They also showed that the position of the band due to $>C{=}C<$ between 1650 and 1500 cm^{-1} correlated with the length of the sequence. Subsequent workers [48–51] have used these bands to study similar sequences in various polymers.

6.4.3 Quantitative Polymer Studies

The applications described have largely dealt with the physical and chemical characterization of polymers. In several of the cases cited, quantitative aspects can also be measured. Quantitative measurements in Raman spectroscopy of polymers vary from the relatively simple to the quite complex. The relative intensities within a normal Raman spectrum will simplistically be directly proportional to concentrations of the species present, the laser power and the Raman scattering cross-section. As the scattering cross-section is very difficult to determine, absolute band strengths are rarely, if ever, determined. Determination of relative strengths by using band ratios is most common. This method can easily be employed in the examples already cited of PVC degradation, degrees of polymerization of acrylics and epoxides, and filler content. More complex studies often have to employ more sophisticated quantitative techniques which involve several bands or complete regions of the spectrum. These can employ principal component analysis (PCA), factor analysis (FA), principal component regression (PCR) and partial least squares (PLS). This is particularly the case where multi-component blends, composites or morphological features are being studied. An example of this is the modelling of PET density from normalized, mean-centred FT Raman spectroscopy studies. The bandwidth of the carbonyl band in the Raman spectrum has been associated with the density and hence the sample crystallinity [52].

6.5 COLOUR

6.5.1 Raman Colour Probes

Raman spectroscopy is very sensitive to coloured molecules. Indeed using visible laser sources highly coloured materials can exhibit too much sensitivity and be the cause of many difficulties. Coloured molecules can be so strongly absorbing as to thermally degrade. There are ways of diminishing this effect by spinning samples or diluting them in a matrix such as hydrocarbon oil or potassium bromide (KBr) powder. Moving to a higher wavelength source can mitigate this effect but even small amounts of some colours can still cause problems. A combination of two Raman techniques, described in Chapter 5, can both overcome the fluorescence effects and show large increases. Surface enhanced Raman spectroscopy (SERS) gives a 10^6 increase in sensitivity and also strongly quenches fluorescence. Resonance Raman spectroscopy (RR) gives a 10^4 increase. The combination of the two techniques, surface enhanced resonance Raman spectroscopy (SERRS), gives a $\sim 10^{10}$ enhancement. Early SERRS studies of rhodamines 3G, 6G (Figure 6.11) and 3B (**1**) have detected the dye in solution at $\sim 10^{-10}$ mol but more recent studies have claimed detection limits of $\sim 10^{-18}$ mol which is roughly equivalent to having 35 molecules in the beam at a given time [53–56].

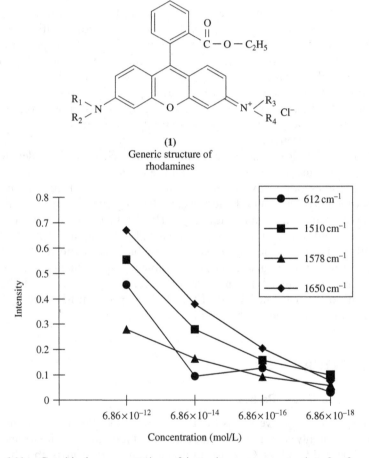

(**1**)
Generic structure of
rhodamines

Figure 6.11. Graphical representation of intensity vs. concentration for four peaks selected from the R 6G SERRS spectra using nitric acid aggregation (points corresponding to R 6G concentration of 6.8×10^{-10} M were omitted). (Reproduced with permission from C. Rodger, W.E. Smith, G. Dent and M. Edmondson, *J. Chem. Soc. Dalton.*, 791 (1996).)

The experiment can be carried out in a laboratory only in the absence of a dye laser spectrometer as this is a potential source of contamination at these levels. This low level of detection limit has opened up the field of SERRS probes to otherwise prohibitive areas of vibrational spectroscopy. This is particularly true in the biological area. In many cases the applications are outside the scope of vibrational spectroscopy. Infrared spectroscopy is limited by the strong absorption of water. Conversely, Raman spectroscopy can cope with aqueous media and the applications are expanding. One area of much interest is tagging DNA with a chromophore rather than with a fluorophore (see Section 5.7). By employing SERRS techniques which are very sensitive and specific to the chromophore, levels of DNA at 10^{-15} molar have been detected [57–59].

6.5.2   *In Situ* Analysis

The major advantage of Raman spectroscopy in the application of colour probes is that vibrational spectroscopy can be used as a largely non-destructive, *in situ* sampling technique. Dyes and pigments can have the spectra recorded as neat samples inside vials and bottles with no concern for sample preparation changing the form. The same dyes and pigments can now be studied while in use in the applications for which they were designed. The strength of Raman signals from coloured molecules has already been discussed. However when the dye and pigments are used to colour materials, the amount of dye is relatively low, e.g. a 2% dye loading on a polymer, can still be detected by Raman spectroscopy [60]. The low concentration has the effect of diluting the colour and reducing thermal degradation, similar to the mull and halide disk techniques. This enables strong Raman spectra to be recorded where the bands due to the colour component can predominate. The strong bands which appear in the Raman spectrum are from the part of the molecule generating the chromophore and weak bands appear from the rest of the molecule. The reason for these extra-strong bands with visible laser sources is quite often attributed to resonance. This does not explain why a similar chromophore enhancement is sometimes observed with 1064 nm laser sources on dyes with only weak absorptions in this region. The likelihood is that enhancement is due to pre-resonance [61] (see Section 4.2.2). This feature can be used to study changes in dye conformers and the effect or otherwise on the chromophore part of the molecule not only on the neat dye but also that dispersed in the material being dyed. A disperse red dye (2) for polymer textiles has been shown to exhibit more than one physical form by liquid and solid NMR studies [62, 63]. NIR FT Raman spectra of the solid samples of the dye showed differences in the azo band positions and also in the backbone structure of the molecules. A piece of poly(ethylene terephthalate) ester cloth dyed at $\sim$2% level with disperse red was examined directly in the spectrometer beam at 1064 nm excitation. The spectra of forms I and II are shown in Figure 6.12a. The dyed cloth and undyed cloth are shown in Figure 6.12b.

The bottom spectrum of Figure 6.12b whilst showing strong bands due to the poly(ethylene terephthalate) fibres also shows bands due to the dye. It is clear

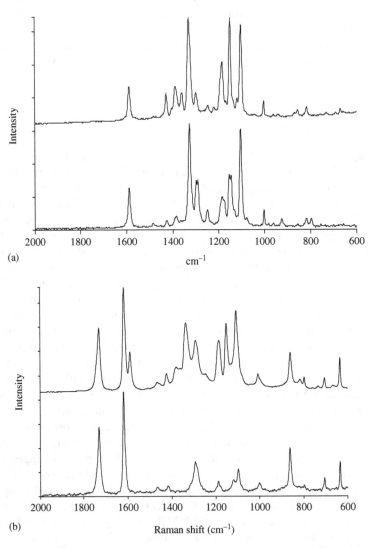

Figure 6.12. NIR FT Raman spectra of (a) azo dye, forms I (top) and II (foot); (b) red cloth – dyed (top); PET (bottom).

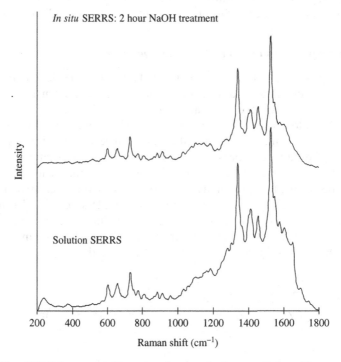

O$_2$N — [ring] — N≡N — [ring] — N(CH$_2$CH$_2$COOCH$_2$)$_2$

NHCOOCH$_3$

(2)

Disperse Red

from this spectrum that form I of the dye is predominant whilst the dye is in the fibre.

The resonance effect of the chromophore can be enhanced by use of the SERRS technique. It also enables *in situ* identification of chromophores from small samples. This technique has been used to identify chromophores in pen inks [64], dyed fibres [65] and lipsticks [66]. With polymer fibres the dyes are dispersed in the fibre but with cellulose the dyes can be reacted onto the fibre. The SERRS technique has been used to detect dyes at very low levels reacted onto the fibre [68] as shown in Figure 6.13. The lower spectrum shows the dye

In situ SERRS: 2 hour NaOH treatment

Solution SERRS

Intensity

200   400   600   800   1000   1200   1400   1600   1800

Raman shift (cm^{-1})

Figure 6.13. SERRS spectra of reactive dye attached to a cellulose fibre and in solution. (Reproduced from reference [68] by kind permission of Dr C. Rodger.)

in solution. The upper spectrum is of the dye attached to a fibre which had been treated in caustic solution for 2 h. This was to ensure that the dye was firmly attached and would not leech back into the SERRS colloid.

Dyes and pigments are used for printing in a number of different printing mechanisms and applications [67]. One of the fastest growing areas for the use of dyes is in the ink of inkjet printers. The dyes first developed for these inks were very similar to those used for textiles. The requirements of the ink manufacturers were very different from the textile dyers. Only three main colours, cyan, yellow and magenta were required. The dyes had to have high colour, with good light and wet fastness. Unlike textiles the dyes could not be fixed to the paper by boiling! Dye development occurred along the non-chromophoric part of the dye to increase fastness to the inkjet media. Originally, the inkjet medium was cellulose paper of varying complex composition and pH, but now it can be polymers, gels, textiles or electronic surfaces. Studies of the behaviour of the dyes on various types of paper surfaces to determine the effects on the chromophore of pH have been carried out [68, 69]. By using both SERRS and NIR FT Raman techniques, the behaviour of the dyes can be studied as solids, on the surface of the paper and below the surface. The effect of varying the non-chromophoric constituents on the chromophore in various parts of the media under differing pH conditions can be studied in this way.

6.5.3   Raman Studies of Tautomerism in Azo Dyes

The changes seen in the *in situ* analysis of dyes can be due to physical or chemical form changes. Azo dyes are amongst the most extensively employed dyes in the colour industry. They are used for their electrical properties as well as their wide colour range. The azo group (3), which simplistically exhibits azo-hydrazo tautomerism, is symmetrical and therefore very weak in the infrared spectrum. The band can be strong in Raman spectra [70] at approximately $1450\,\text{cm}^{-1}$. When the hydrazo is formed, the $-C{=}N-$ band can be seen in the Raman spectrum at approximately $1605\,\text{cm}^{-1}$ with another strong band at $\sim 1380\,\text{cm}^{-1}$, the origin of which is still not fully understood. Whilst this group is very important in colour chemistry, being a significant component in many dyes [71–73], the detection and interpretation of these bands can be complex and have been the subject of many studies by the authors.

(3)

Azo-hydrazo tautomers

A simple example of the comparative information obtained from the infrared and Raman spectra is a yellow diazo dye (**4**). The dye has a generic structure with azo, carboxyl and triazine groups. The upper infrared spectrum in Figure 6.14 is complex due to strong hydrogen bonding between groups. Bands at $3500-2000\,cm^{-1}$ show a mixture of salt and free carboxylic acid groups. The band at $\sim1550\,cm^{-1}$ is probably due to the triazine ring. The general broadness is due to both the hydrogen bonding effects and the large size of the molecule. The azo groups cannot be seen in this infrared spectrum.

a(HOOC)

(COOH)b

a = 1 or 2
b = 1 or 2
X = morpholinyl or NHR

(**4**)

Generic structure of yellow diazo

Conversely, in the lower Raman spectrum the azo and aromatic bands dominate the spectrum, with the hydrogen bonded groups being too weak to observe.

6.5.4 Polymorphism in Dyes

In the azo dyes, chemical group changes which could affect the colour properties were studied. Physical changes can also take place in the molecular structure

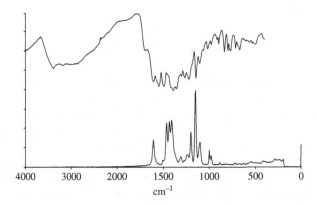

Figure 6.14. Infrared and NIR FT Raman spectrum of azo dye. (Reproduced with permission from J. Chalmers and P. Griffiths (eds), *Handbook of Vibrational Spectroscopy*, vol. 4, John Wiley & Sons, Inc., New York, 2001.)

of a dye which in turn can also affect the properties. This effect is known as polymorphism, described in Section 6.7. In the electrophotography industry the electrical properties of organic materials can change the effectiveness of the device. In a photocopying machine or laser printer one stage of copying the image requires a charge to be generated in a photoreceptor which usually comprises several layers. The base is usually conducting, above which is a charge transfer generation (CTG) layer. On exposure to the laser an electronic hole is generated. This hole is transferred to the surface of the photoreceptor via charge transport material (CTM) in the charge transport layer. The wavelength of the illuminating laser is about 750 nm. The charge generating layer is less than 1 μm thick to allow mobility and transfer of the generated hole. Dyes are often used to absorb energy at the irradiating laser wavelength and compose most of the CTG layer. The most adaptable dyes are the phthalocyanines. Their electrical properties vary with the metal co-ordinated in the ring. Metal-free phthalocyanine (5) also has interesting electrical properties. The position of the ring breathing band at $\sim 1540-1510\,cm^{-1}$ in the Raman spectrum is quite characteristic as can be seen in Figure 6.15.

(5)

Metal-free phthalocyanine

Titanium phthalocyanine (TiOPc) is most widely used. However, the speed of charge generation is vital. TiOPc has several polymorphs and only one is suitable for optimum charge generation. The Raman spectra of various TiOPc polymorphs in Figure 6.16 show subtle but distinctive changes. Type IV has a very small extra band at $765\,cm^{-1}$. The spectrum of a drum placed directly in the beam of an NIR FT Raman spectrometer (Figure 6.17) shows this feature quite clearly. Bands are present from the other components – the covering polycarbonate layer, the CTM component and the base layer of titanium dioxide – but none have this band in their spectra. The confirmation of the correlation of this distinctive feature comes from the relative charge transfer speed of the devices with and without form IV.

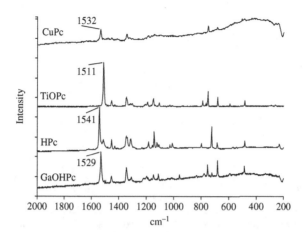

Figure 6.15. NIR FT Raman spectra of various metal phthalocyanines. (Reproduced from NIR FT Raman examination phthalocyanines at 1064 nm, G. Dent and F. Farrell, *Spectrochim. Acta*, 1997, 53A, 1, 21 © 1997 by kind permission of Elsevier Science-NL, Sara Burgerhartstraat 25, 1055 KV Amsterdam, The Netherlands.)

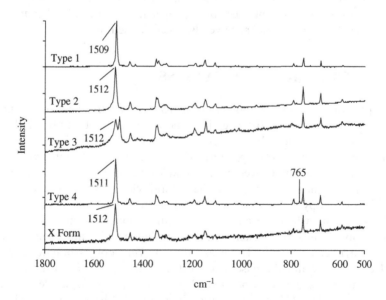

Figure 6.16. NIR FT Raman spectra of the polymorphs of TiOPc. (Reproduced from NIR FT Raman examination phthalocyanines at 1064 nm, G. Dent and F. Farrell, *Spectrochim. Acta*, 1997, 53A, 1, 21 © 1997 by kind permission of Elsevier Science-NL, Sara Burgerhartstraat 25, 1055 KV Amsterdam, The Netherlands.)

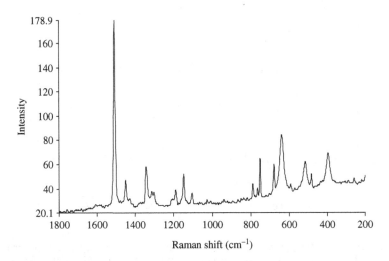

Figure 6.17. NIR FT Raman spectra of photocopier drum with anatase TiO_2. (J. Chalmers and P. Griffiths (eds), *Handbook of Vibrational Spectroscopy*, vol. 4, John Wiley & Sons, Inc., New York, 2001.)

In this application a coloured molecule has been used and studied by Raman spectroscopy because of the electronic properties. Studies of other materials used in the electronics industry have also been carried out.

6.6   ELECTRONICS APPLICATIONS

In the previous section the application of Raman spectroscopy to colour was generally concerned with the chemical and/or physical structure of the dye. However, colour can also be used effectively in electronic devices, for example to provide a detector or to produce an image. In electrophotography dyes, including phthalocyanine dyes were found to be very useful. The selectivity of resonance Raman scattering and the natural selectivity of Raman scattering make spectroscopy a good probe for these materials. As described in the previous section, when the charge generating drum is examined by Raman spectroscopy, bands are present from the phthalocyanine and from the other non-coloured components. These are the covering polycarbonate layer, the CTM component and the base layer of titanium dioxide. Information on all these components can be obtained. The titanium dioxide is clearly in the anatase form. The CTM composition tends to be based on triarylamines (**6**) with extra-strong aromatic bands due to the multiple conjugation. These materials transfer electrons under charge and are therefore strongly polarizable. The positions of these bands are very sensitive to the environment. These

(6)

Triarylamine

types of material besides being used in bulk, in copier and printer drums, are also being employed as very thin films or coatings in organic semiconductors such as field effect transistors [74–76], solar cells [77] and light emitting displays [78, 79]. The ability to carry out microscopic or *in situ* examination allows the study of these materials in the devices in which they are used. The types of materials employed in these devices are usually conducting polymers with conjugated π-electrons. Typical polymers initially employed were polyacetylene, poly(p-phenylenes), polythiophenes and poly(triarylamines) [80–83]. Several of these have been available from commerial sources since 1996 [84]. Poly(p-phenylene vinylene)s (PVPPs) and spiro compounds have also been developed [85, 86]. A typical band gap of these polymers is 1–3 eV. A conjugated polymer can be doped with electron acceptors such as halogens, or donors such as alkali metals. The polymers can be discussed in concepts, new to the organic chemists, of solitons [87], polarons [88] and bipolarons [89]. Essentially the polymers carry charges or emitting light. As in other polymer applications the effectiveness of these materials is dependent on the morphology as well as the chemistry. However a further advantage is the change in the spectra that takes place in the excited state of the polymer. The electronic behaviour of the building block monomers, on which the polymers are based, can be studied by cyclic voltammetry (CV) to determine the oxidation and reduction points. This leads to Raman spectroelectrochemistry whereby the Raman spectra of the materials can be measured during the voltage cycle. This in turn leads to an understanding of the spectra generated in a device. Whilst this may appear to be a recent development, the original work which led to the discovery of the SERS effect by Fleischmann and Hendra [90] was of course an attempt to study electrochemical changes by Raman spectroscopy. Raman spectroelectrochemistry is discussed further in Section 7.1

Whilst electronically active polymers and organic semiconductors are at the forefront of technology, more conventional semiconductors have also been studied by Raman spectroscopy. Elements such as carbon, germanium and silicon are widely used in the electronics industries for both their mechanical and electronic properties. As discussed in Section 6.9 this is probably one of the largest use for Raman spectroscopy as a quality control tool. Stress measurements have been discussed in polymers but can also be carried out in electronic devices, just as in polymers, shifts occur in the wavenumber position of the Si Raman bands. This shift can be used to obtain Raman maps of stress and crystallinity in silicon wafers [91, 92]. In heavily boron-doped silicon the LO (longitudinal optical) phonon line, which is a very sensitive Raman band, shifts to lower frequency and broadens as the boron concentration increases [93]. Films of fluorine-doped silicon dioxide have been monitored by Raman spectroscopy to determine the fluorine to oxygen ratios and hence the dielectric properties [94]. Silicon crystals are employed in ultra large-scale integrated (ULSI) circuits [95]. The Raman spectra of the crystals give information on the lattice modes and the behaviour of the crystals. The vibrations of crystals used in semiconductors are reflected in the behaviour of phonons. It is this behaviour which is studied by Raman spectroscopy. The structural characterization can include the crystallinity, crystallographic orientation, superlattices of mixed crystals, defects, and stacking faults. Besides the structural characterization, electronic characterization can be carried out. Both bound and free charges can contribute to Raman scattering, through collective and single-particle excitation processes. For those interested in studying the processes in greater depth than can be described here, there is the series 'Light Scattering in Solids I–VII' which appeared between 1982 and 2000 in *Topics in Applied Physics* published by Springer in Berlin. There are several other excellent review articles as well [96–99].

6.7 BIOLOGICAL AND PHARMACEUTICAL APPLICATIONS

6.7.1 Introduction

The biological and pharmaceutical areas of application lend themselves well to investigations by Raman spectroscopy. There are two particular aspects which are common to both areas. One is the possibility of studying materials *in situ*. In biological systems, wet live cells have been studied whilst in the pharmaceutical area tablets inside polymer containers have been identified. Another use of Raman spectroscopy is to study physical structure. In particular polymorphism is very important in the pharmaceuticals industry and there are a number of Raman studies. Similarly, in proteins and peptides, secondary structure has been studied at length, though in this case interpretation of the results can be quite complex. Some include more sophisticated techniques such as Raman optical activity (ROA) (see Chapter 7). This topic covers a very wide area of application on which specific chapters and books have been written. Indeed Chapters 4, 5 and 7 have a number of examples of Raman spectroscopy being used very effectively for biological applications. In this section we can only give a flavour of the vast area of research and application. References and a bibliography will assist those with deeper interest.

6.7.2 Biological Applications

Raman spectroscopy has advantages and disadvantages for the study of biological molecules. Amongst the advantages is the ability to study *in situ* aqueous systems. One of the disadvantages is that in using highly focussed beams, sensitive tissue is easily damaged. However there is a very wide range of biological systems which can be and have been studied by Raman spectroscopy [100]. As polar groups such as carbonyls, amines and amides are weak in the Raman spectrum, this would appear to be a disadvantage. However groups such as –S–S–, –SH, –CN, –C=C– and aromatic rings give strong distinctive bands for specific characterization. Other groups such as carbonates and phosphates can also have distinctive Raman bands. The technique can also detect changes in physical form, so polymorphism, secondary structure in peptides and general molecular backbone changes can be detected. The use of 180° scattering with the development of microscopes and microprobes has led to automation for fast screening and the ability to examine systems *in situ*. In the case of microprobes, the enhancement techniques such as SERS and SERRS which can use metal colloids as part of the probe have brought large increases in sensitivity. The authors have used these techniques, as described in other sections, to detect DNA, and along with others are approaching single molecule detection. Among a vast number of applications reported are binding studies, genomics, lab on a chip, proteomics, protein interactions and solid phase synthesis. DNA and protein arrays as well as cell growth have been studied. There are also a large number of published studies in the food and biomedical fields. A few examples are: transitions in amino acid crystals [101], single-cell bacteria [102], bacterial spores [103], carotenoids in numerous systems including atlantic salmon [104], characterization of microorganisms [105], fungi [106], grain composition [107], liposome complexes [108], yeast [109] and benign and malignant tissue in thyroid [110] and human breast tissues [111]. Specific studies of chromophore containing proteins can give good structural information from suspensions as illustrated for P450 enzymes in Section 4.4.2. Gradually, as equipment and skill improves through the work of some excellent groups, Raman scattering is becoming a useful technique in medicine, detecting *in situ* species such as pre-cancerous tissue and plaque.

6.7.3 Solid Phase Organic Synthesis/Combinatorial Chemistry

As mentioned in the previous section, one of the growing areas of interest for Raman spectroscopy is solid phase organic chemistry (SPOC) or combinatorial chemistry. Rather than being carried out in solution, reactions take place in/on

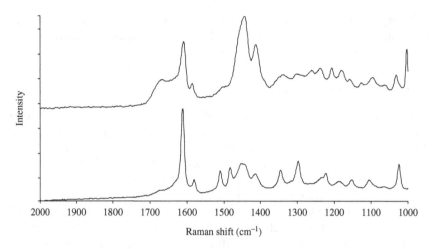

Figure 6.18.   Spectrum of acrylate bead (bottom) and with peptide (top).

solid supports. These supports are usually beads of polyacrylate or polystyrene with reactive end groups such as hydroxyls, chlorine or glycol groups [112]. The beads are porous and have a cellular structure. The beads swell in the solvent and can become gelatinous. The reactants diffuse in and out of the beads with the reactions taking place at the active sites. Analysis of the beads is dependent on the information required (Figure 6.18). Often the phrase "on bead analysis" is used which leads to the erroneous conclusion that surface techniques apply. However, Raman spectroscopy can penetrate into the bead, enabling effective signals to be obtained. Approximately 90% of the reaction takes place in the bead. For research purposes single beads are often monitored but for more routine analysis batches of beads are usually studied. A recent comparison [113] of FTIR and FT Raman methods highlighted the strengths and weaknesses of both approaches. The advantages of Raman spectroscopy is once again the lack of sample preparation, weak absorptions from the solvents and the ability to carry out *in situ* studies. The beads themselves can be studied during preparation as reactive beads, or reactions taking place in/on the beads can be studied.

Peptide reactions can have several sequences of protection and de-protection. The 9-fluorenyl-methoxy-carbonyl (Fmoc) **(7)** strategy is an early developed system [114, 115].

(7)

Fmoc-Cl

The stages have been monitored directly on the beads by investigating changes in the secondary structure [116]. The amide I and III bands were studied to gain information on the secondary structure of the growing peptide chain. Comparative studies of *in situ* reactions on beads have been carried out by FTIR and FT Raman spectroscopy [117–119]. More recently the use of dispersive Raman spectrometers has also been investigated for studies with flow through cells [120].

6.7.4 Pharmaceuticals

The advantages of Raman spectroscopy to the pharmaceutical community come largely from the ease of use, minimal sample handling and strong differences in relative scattering strengths of packaging materials, tablet excipients and the active agents. These strengths combined with the use of microscopes and fibre optics have seen a large growth of use in the pharmaceutical industry. An early worker with FT Raman quickly recognized the advantages and opportunities in the pharmaceutical industry [121]. A review by Cutmore and Skett [122] is still very valid. Although the fluorescence issue is still a problem with dispersive instruments, applications such as drug screening and polymorphism have been studied by both FT and dispersive techniques. In the area of fibre-optic coupling, microprobes and imaging dispersive technology is very much at the forefront. As with other application areas in this chapter, there is not room for a comprehensive review but a few illustrative examples are given.

☐ NON-CONTACT *IN SITU* MEASUREMENT

For quality control of manufacturing and formulation the ability to check directly inside a polymer package produces tremendous time and cost savings. Imaging of tablets can be carried out to check the distribution and relative amounts of active agent, additives and binders present. The active drug is often an aromatic-based compound with distinctive Raman spectra whilst the other components are sugar, cellulose or inorganic-based materials. The active component itself can also have variable properties dependent on the physical form or crystallinity. These can affect dissolution rates and hence the efficacy of the drug. Drug samples, including both prescription drugs and drugs of abuse, can be measured *in situ* in clear plastic wrappings. This can be important both in the speed of analysis and in preventing sample contamination. If a microscope system is used, the laser beam can be focussed onto the surface of the tablet through the plastic packing material. The high-power-density area created provides most of the Raman scattering and therefore discriminates in favour of the tablet or powder. In addition, the scattering from the drug is usually relatively intense compared to that from the plastic material. In Figure 6.19, an example of this type of experiment is shown.

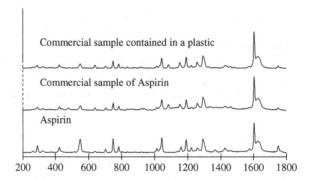

Figure 6.19. Raman spectra of aspirin, commercial sample of aspirin, and aspirin inside plastic wrapper.

A commercial aspirin tablet, an identical tablet wrapped in plastic and laboratory-synthesized aspirin powder all give almost identical spectra with the same accumulation times. As can be seen, the spectra of the aspirin can clearly be identified in each case. There are relative band strength differences. In the spectra the aspirin gives strong bands, the binder gives weak bands and none are present from the wrapping material. This demonstrates significant selectivity in Raman scattering experiments. This simple experiment was carried out using 30-s accumulations in a standard Raman spectrometer. It indicates the simplicity with which a compound can be identified *in situ* and provides significant information which helps both to identify a specific compound and to indicate impurities. For example, the inorganic filler in the tablet gives rise to a band at about 1100 cm^{-1} which is not present in the laboratory aspirin sample and the additional band at about 1370 cm^{-1} shows that the laboratory aspirin sample was not pure.

❐ MOLECULAR SPECIFICITY

Raman scattering can be obtained from a set of drugs of abuse. Each spectrum is molecularly specific. Initial identification of a sample without the matrix is very simple by Raman scattering. However in real world samples the drugs are often in a matrix of several compounds. This matrix can cause fluorescence which swamps the image; in the case of inorganics, Raman spectroscopy can be used to identify the impurity.

❐ POLYMORPHISM

One property which can greatly affect the efficacy of a drug is polymorphism. The term can be used in the biological and pharmaceutical worlds with totally different meanings. In most of the chemistry applications and here in pharmaceutical applications, McCrone's definition [123] is used to mean differing physical forms of the same molecule. Raman spectroscopy is ideally suited to studying polymorphism as the lack of sample handling minimizes the risk of converting the form during measurement as can occur with other techniques. Surprisingly, an extensive review by Threlfall [124] of analytical techniques employed to study polymorphism did not have a very large section on Raman spectroscopy. Much of what was reported was with FT Raman. Even so the technique has shown differences with polymorphs of several compounds [125–128] including the B and C forms of naphthazarin [129] and the morphological composition of cimetidine [130]. In addition to that of the active agent, the differing crystallinity of excipients such as glucose have also been recorded [131]. The strict confidentiality surrounding pharmaceutical drug structures are the composition of tablets and this means that much of this work has not been reported in the open literature. For example, a recent application note [132] on polymorphism by dispersive Raman refers only to forms A and B without identifying the drug. A review by Frank [133] contains many examples of the use of Raman spectroscopy in the study of pharmaceuticals.

Yet another example of an unnamed pharmaceutical intermediate is shown in Figure 6.20 recorded by the authors. The FT Raman spectra were recorded

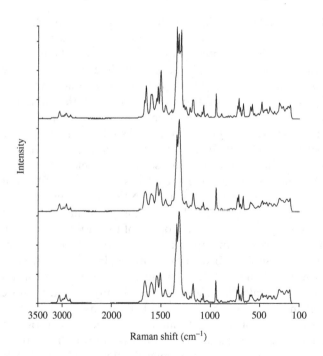

Figure 6.20. Crystallinity in pharmaceutical intermediate: crystaline form (top); amorphous (middle); mixture of crystalline and amorphous forms (bottom).

of pure crystalline and amorphous forms of the intermediate. A suspect batch of the amorphous form was examined without opening the sampling vial. The spectrum shows evidence of crystallinity at $\sim 1500\,\text{cm}^{-1}$. However some examples of the use of Raman spectroscopy with pharmaceuticals for which the structure is known have been published. For example, the use of Raman scattering for high throughput screening of carbamazepine (**8**) has been reported by the Novartis company [134].

(8)

Carbamazepine

6.8   FORENSIC APPLICATIONS

Modern developments in equipment for the detection of Raman scattering make the method very useful in the area of forensic science [135]. The main advantages of Raman scattering in forensic science are the non-invasive, non-contact nature of the method, the ready coupling of the instrumentation to microscopes so that detection of very small amounts is achieved easily, and the molecularly specific nature of Raman scattering. Both visible and NIR FT systems are applicable. In forensic science, it is usual that these are coupled either to a microscope or to a fibre-optic head so that sampling of small areas of material can be performed. In current applications, the microscope is more widely used, but the requirement to use the instrument in the field may, in certain circumstances, dictate the use of a simple, small fibre-optic-coupled head. The main advantage of using the NIR system in forensic science is reduced fluorescence, which can be a major problem with a sample in the matrix but the simpler, more flexible optics of the visible system can make this the system of choice in many applications. To this extent the advantages are very similar to those described in Section 6.3. Indeed it can be argued that forensic science is already applying those concepts described in previous sections on the micro-scale. Specific examples already quoted in this chapter are drugs, fibres and colour probes in several situations. Indeed it can be argued that the advancement of Raman technology has been a contributing factor in enabling analytical scientists to take a forensic approach to problem solving. Besides fibre identification and biological effects, where SER(R)S can be advantageous, one of the areas of significant interest to forensic science is the identification of explosives. Nitro-group containing RDX and PETN are components of plastic explosives that have very low vapour pressure. These can be identified readily from one small particle by Raman scatering. It is possible to examine a surface such as that from a fingerprint to obtain either an image of the surface under the microscope or to map a larger area. Mapping enables the analysis of larger areas. Images of a fingerprint showing the presence of a particle of RDX have been recorded [136].

6.9 PLANT CONTROL AND REACTION FOLLOWING

6.9.1 Introduction

Whilst many applications mentioned in this chapter so far have used static, or *in situ* measurements, a growing area of interest for Raman spectroscopy is reaction following. In principle the technique is ideal, being non-invasive, able to detect from within glass vessels and aqueous media, and able to carry out monitoring at long distances. Published reports of industrial applications have until recently been relatively sparse. Many have covered laboratory trials or proof of principle. This has been due to several reasons. Fluorescence makes the technique very application-specific. If an instrument is developed to monitor a specific reaction, it does not follow that it can be easily transferred to other applications. The instrumentation was until recently very large, expensive, required specific environmental conditions, i.e. dark rooms, laser interlocked doors. With the introduction of modern variable filter instruments and FT spectrometers, the instrumentation has become much more user friendly and adaptable. Flexibility of application has increased with lasers sources at 785 and 1064 nm and in the UV, thus reducing fluorescence for reaction following. The portability and increased simplicity of the instruments has also opened up the possibility of on-plant monitoring. The latter is somewhat under-reported. This tends to be due to companies wishing to maintain the commercial edge gained from greater efficiencies achieved by tighter plant control. The number of reported examples of reaction following and plant monitoring is increasing. Two excellent reviews have been published [137, 138] on the parameters to be aware of and potential pitfalls in introducing Raman spectroscopy in an industrial plant. A few typical applications are given.

6.9.2 Electronics and Semiconductors

Probably one of the biggest quality control (QC) applications [139, 140] for Raman spectroscopy is monitoring of the protective diamond-like films (DLF) for computer hard disks. Information on hydrogen content, sp^2/sp^3 ratios and long-range ordering is available from the Raman spectra. Instruments solely dedicated to these measurements can automatically predict the tribological qualities of the films [138, 141]. QC techniques have been established for the physical and chemical characterization of semiconductors, which can include crystal size and form, dopant levels, stress and strain. This is a major topic in itself for which there are several reviews [142–144] and which is also discussed in Section 6.6. Another feature of process monitoring in this industry is in the deposition and/or growth of thin films. Raman spectroscopy is particularly applicable when the process occurs under vacuum or at elevated temperatures. This could be potentially important for monitoring novel heterostructures [145]. The growth of InSb on Sb(1 1 1) has been studied [146] with thicknesses of 0–40 nm. The growth of ZnSe on GaAs at 300 °C followed by capping with Se has been reported [147]. This was followed by crystallization studies of the Se layer. Other studies [148] have included the nitridation of ZnSe on the GaAs and the growth of CdS on InP(1 0 0). This industry probably makes the largest use of process and QC applications of Raman spectroscopy. It is an application which is extremely important to modern life and technology and yet does not greatly feature in conventional literature on analytical chemistry.

6.9.3 PCl₃ Production Monitoring

If a plant analyst was asked which reactions would cause the largest problems in monitoring, then elemental phosphorous combined with chlorine in a boiling liquid would feature high on the list. Yet this, the production of PCl_3, is one of the best known examples of Raman on-line monitoring [149] developed by Freeman *et al.* The levels of chlorine have to be maintained to prevent the production of PCl_5. Using a sampling loop, an FT Raman spectrometer and fibre bundles with laser powers of 2 W, and with 140 scans and $16\,cm^{-1}$ resolution detection levels of <1% for P_4 and Cl_2 were attained. Side products of $POCl_3$ and $SbCl_3$ were also monitored but these degraded under the high laser power. Gervasio and Pelletier [150] refined the measurement with a CCD-dispersive system, a 785 nm laser and a direct insertion probe.

6.9.4 Anatase and Rutile Forms of Titanium Dioxide

One of the earliest publications on plant control is the monitoring of the physical form of titanium dioxide, a very bright and white commonly used pigment. However the pigment exists in different physical forms. The major ones being anatase and rutile. The rutile form, having a higher opacity, is more commonly used. Both forms have very distinctive bands in the Raman spectrum (see Figure 6.3). The anatase form has bands at 640, 515, 395 and $145\,cm^{-1}$ whilst the rutile bands appear at 610 and $450\,cm^{-1}$. These bands have been used to quantitatively measure 1% of anatase in rutile with enough accuracy for semi-automated plant control [151]. The greatest difficulty with this measurement was the dusty environment inside the production plant which resulted in major engineering problems. These problems and their solution have been colourfully described at length by Everall *et al.* [137].

6.9.5 Polymers and Emulsions

As described in Section 6.4, one of the simplest reactions to follow by Raman spectroscopy is the loss of the $>C=C<$ bond. This gives very clear and strong bands in simple monomers such as acrylates, vinyl acetates and styrene. There are numerous literature references covering a range of applications from the simple to the complex. Styrene (S) and methylmethacrylate (MMA) have been studied in a reaction cell [152], homopolymerizations of MMA and butyl acrylate (BuA) have been monitored by FT Raman in laboratory reactors [153] and an analysis of a complex quaternary polymer (S/BuA/MMA/cross-linker) has been carried out [154] to show the consumption of monomers followed by the composition of the resultant copolymer. It would be easy to think that the $>C=C<$ bond band could be monitored quantitatively directly and this can indeed be the case [152, 153]. However, changes in laser intensity, spectrometer response and inhomogeneity can lead to the band having to be normalized. This is usually carried out by choosing a band not affected by the reaction being followed. Unfortunately this simplistic approach cannot be used.

Several workers [152, 155] have reported intensity changes in bands, used for reference purposes, between the monomer and polymer states. This has been discussed in depth by Everall [156]. Whilst these systems can be monitored quantitatively, similar reactions taking place in emulsions require even greater care. The monomer can exist as droplets, dissolved in water, as micelles or in the polymer phase. The band intensity and wavenumber position can be affected by which phase the monomer is in. Temperature and pressure changes can affect both the spectrum and the phase in which the monomer resides. Equally importantly, the detection of the drops can be affected by the size relative to the wavelength of the laser exciting line (see particle size effects, Chapter 2). These factors need to be taken into consideration but should not prevent analysis taking place, as a published application [157] monitoring latex emulsion polymerization has shown. Vinyl and acrylate monomers have been largely described here but other systems have also been reported such as cyanate esters [158], epoxies [159], melamine-formaldehydes [160], polyimides [161] and polyurethanes [162]. All the applications described so far have taken place largely in bulk in reaction vessels. Raman spectroscopy has developed in these applications through the use of direct probes and/or coupling with fibre optics which in the case of visible laser sources can be monitored at distances of up to several metres. Another area of application employing *in situ* analysis by Raman spectroscopy is in extruders, fibres and films. In these cases the physical properties of the polymer can be studied as well as the chemical composition. Early work by Hendra [163], taking advantage of polarized Raman spectroscopy, analysed polymers in a lab-scale extruder for information on crystallinity, chain conformation and orientation. Modern instrumentation and fibre optics developments have enabled similar measurements to be carried out *in situ* on pilot plant and full scale production facilities. Recently Chase has also taken advantage of polarized Raman spectroscopy to study fibres at varying points of the drawing process and has carried out extensive studies of fibres on spinning rigs [164, 165].

Similar measurements would appear to be applicable to polymer film production. The measurements are quite complex as the morphological properties can develop in three dimensions. Farquharson and Simpson [166] demonstrated the feasibility with a dispersive spectrometer and a 5 m fibre bundle in a first-reported Raman on-line analysis of polymer film production. Since then Everall has carried out extensive work on the composition of polyester film on a moving production line using imaging fibre probes coupled though 100 m of fibre [167, 168]. One of the features discovered in this work was the difficulty in handling fluorescence in a moving film compared to static measurements. When a polymer film is static in the beam, low levels of fluorescence can be burnt out (photobleached). With a moving film the sample is rapidly refreshed maintaining the level of fluorescence. Moving from visible excitation towards 785 nm can reduce these effects but Everall found that addition of reclaimed polymer to the stream was a cause of fluorescence [156]. Monitoring polymer film of various types is an application of plant control where Raman spectroscopy would have been expected to have wide use. The lack of literature could be due to commercial sensitivity.

6.9.6 Pharmaceutical Industry

The pharmaceutical industry has many potential applications as described in Section 6.7. The minimal sample handling and the ability to 'see' into polymer containers would be expected to lead to numerous QC applications. Instrument manufacturers claimed expanding sales due to interest in polymorphism alone. Yet little literature, apart from manufacturers' application notes, is available for on-plant applications. Again it is probably commercial sensitivity which is the cause. Interested readers should consult the later references in Section 6.7.

6.9.7 Fermentations

Biotechnology and bioreactors are a fast growing area of technology. The potential advantage of Raman spectroscopy is the ability to study aqueous systems, though fluorescence and particulates are still potential problems. Little was reported prior to the introduction of lasers in the red end of the spectrum. Shaw *et al.* [169] have analysed glucose fermentation with a fibre optic coupled 785 nm laser. They extracted and filtered the liquid to remove yeast cells, which, though not a direct measurement, demonstrated the potential of the technique.

By employing PLS techniques and other modelling techniques, including neural networks, glucose and ethanol contents were predicted with errors of $\sim$4%.

6.9.8 Gases

Raman spectroscopy of gases is not an industrial application which readily springs to mind yet, as stated in Chapter 2, some of the earliest Raman spectroscopy was carried out on gases and vapours in inclusions in minerals and rock. The low sensitivity of Raman spectroscopy to gases, due to the low scattering cross-section and few molecules in a given volume, would appear to reduce the applicability. However these drawbacks have been overcome by using special cells, instruments or remote sensing techniques such as light source techniques similar to radar known as Raman LIDAR [170]. A simple case which demonstrates the sensitivity of Raman spectroscopy to simple molecules is the interference that can occur in weak Raman spectra from fluorescent room lights. Here, sharp emission bands are recorded which can be mistaken for the sample and would certainly cause problems with multivariate analysis routines. The pattern varies with the laser exciting line used. The relatively weak spectra were enhanced by using multipass cells [141] and placing these inside the laser cavity. Quantitative measurements have been made of CO, CO_2, H_2, H_2O, N_2, N_2O, NH_3 and hydrocarbons using a simple spectrometer constructed from simple components and an intracavity gas cell. The Raman analyser, known as a 'Regap' analyser, has been described by de Groot and Rich [171] for measuring and controlling atmospheres in a steel treatment furnace. Large scale QC methods have been devised [172], but methods using Raman microscopes are also used. The applications include accelerator devices for vehicle air bags and degradation of pharmaceuticals inside package products [173]. In each of these applications it is the *in situ* aspects of Raman spectroscopy which overcome the other apparent limitations.

6.9.9 Catalysts

The study of surfaces under an aqueous phase in a glass container may be a nightmare for an infrared spectroscopist but will cause little difficulty for a Raman spectroscopist. At an American Chemical Society (ACS) Conference in 2000, statistics were produced which showed that Raman-based catalysts studies were being published at a rate in excess of 300 a year. Clearly we cannot cover a field of that size adequately in this book but it is well covered in other publications [174]. We will highlight the applicability of Raman spectroscopy in this field and point the interested reader to further material. The simplest advantages have already been stated. Raman spectroscopy can study systems inside vessels over a range of temperatures and pressures. As many catalyst studies are carried out at several hundred degrees centigrade, this is a major factor. Typical studies are in the automotive industry where catalytic converters for vehicle exhausts operate most efficiently at higher temperatures [175]. The crystal structure of metal oxides and metal co-ordination chemistry is a wide field of study. Metals commonly encountered are platinum, palladium, ruthenium, titanium, uranium, vanadium and zirconium. Alumina- and silica-based catalysts are of continuing wide interest. Many of these materials contain a chromophore which opens up the field to resonance Raman studies. This then has the added advantage of increased selective sensitivity. One of the problems regularly encountered in Raman spectroscopy is fluorescence. There has been an increase in the use of UV laser sources for Raman spectroscopy in this area to overcome fluorescence and also to increase sensitivity [176]. Besides complete reaction following, catalytic partial oxidation is important in industry for production of materials such as alcohols. A typical case is the production of methanol from methane [177].

One of the largest areas of study, of course, is in electrochemical reactions, particularly studies on electrode surfaces. In addition to the normal Raman and resonance Raman advantages in this area, the surface enhance effect (SERS) is of great importance. Indeed as previously shown, the SERS effect was first observed on an electrode surface. This coupled to resonance to create the SERRS effect makes Raman a very powerful tool in this area. Very brief references have been made to an ocean of work. The bibliography will point the reader to much more detailed work.

6.10 SUMMARY

This chapter whilst giving only a flavour of the vast range of applications in which Raman spectroscopy has been utilized should lead the reader to realize the specific niche which the technique still occupies. This chapter, together with the previous chapters, shows that, although Raman instruments are not as generic as for some techniques, care in matching the instrument and accessories to a specific application can create a very powerful specific tool which can be of use to both the expert spectroscopist and the general analyst. The next chapter leads into where the technique can yield even more information, albeit requiring, in some cases but not all, expensive, specialist equipment.

REFERENCES

Inorganics and Minerals

1. M. Yoshikawa and N. Nagai, in: *Handbook of Vibrational Spectroscopy*, J. Chalmers and P. Griffiths (eds), vol. 4, John Wiley & Sons, Inc., New York, 2001, pp. 2593–2600.
2. M.S. Dresselhaus, G. Dresselhaus, M.A. Pimenta and P.C. Eklund, in: *Analytical Applications of Raman Spectroscopy*, M.J. Pelletier (ed.), Blackwell Science, Oxford, 1999, pp. 367–434.

3. P.J. Hendra, in: *Modern Techniques in Raman Spectroscopy*, J.J. Laserna (ed.), John Wiley & Sons, Inc., New York, 1996, p. 94.
4. P. Dhamelincourt, F. Wallart, M. LeClerq, A.T. N'Guyon and D.O. Landon, *Anal. Chem.*, **51**, 414A (1979).
5. E.S. Etz, G.J. Rosasco and W.C. Cunningham, in: *Environmental Analysis*, G.W. Ewing (ed.), Academic Press, New York, 1977, p. 295.
6. C. Beny, J.M. Prevosteau and M. Delhaye, *L'actualité chimique*, **April** 49 (1980).
7. C.J. Rosasco, *Proceedings of the 6th International Conference on Raman Spectroscopy*, Heyden, London, 1978.
8. M. Martoja, V.T. Tue and B. Elkaim, *J. Exp. Mar. Bio. Ecol.*, **43**, 251 (1980).
9. *Internet J. Vib. Spectrosc.* [www.ijvs.com].
10. A. Wang, J. Han and L. Guo, *Appl. Spectrosc.*, **48**, 8 (1994).
11. R.A. Nyquist, C.L. Putzig and M.A. Leugers, *IR and Raman Spectral Atlas of Inorganic Compounds and Organic Salts*, Academic Press, 1997.
12. E.L. Varetti and E.J. Baran, *Appl. Spectrosc.*, **48**, 1028 (1994).
13. D.H.M. Edwards and H.J. Schnubel, *Rev. Gemnol.*, **52**, 11 (1977).
14. Y. Kawakami, J. Yamamoto and H. Kagi, *Appl. Spectrosc.*, **57**, 1333–1339 (2003).
15. J. Popp, N. Tarcea, W. Kiefer, M. Hilchenbach, N. Thomas, S. Hofer and T. Stuffler, *Proceedings of the First European Workshop on Exo-/Astr-Biology ESA SP-496*, 2001.
16. R. Frost, T. Kloprogge and J. Schmidt, *Internet J. Vib. Spectrosc.* [www.ijvs.com], 3, 4, 1.
17. D.R. Lombardi, C. Wang, B. Sun, A.W. Fountain III, T.J. Vickers, C.K. Mann, F.R. Reich, J.G. Douglas, B.A. Crawford and F.L. Kohlasch, *Appl. Spectrosc.*, **48**, 875–883 (1994).
18. K. Williams, *Spectroscopy Innovations*, vol. 6, Renishaw Ltd, 2000.
19. D. Fisher and R.A. Spits, *Gems and Gemology*, **Spring** 42 (2000).
20. H.F. Shurvell, L. Rintoul and P.M. Fredericks, *Internet J. Vib. Spectrosc.* [www.ijvs.com], **5**, 5, 2.
21. A. Peipetis, C. Vlattas and C. Galiotis, *J. Raman Spectrosc.*, **27**, 519 (1996).
22. H.G.M. Edwards, M.J. Falk, M.G. Sibley, J. Alvarez-Benedi and F. Rull, *Spectrochim. Acta A*, **54**, 903 (1998).
23. R.J.H. Clark, in: *Handbook of Vibrational Spectroscopy*, J. Chalmers and P. Griffiths (eds), vol. 4, John Wiley & Sons, Inc., New York, 2001, p. 2977.
24. F.R. Perez, H.G.M. Edwards, A. Rivas and L. Drummond, *J. Raman Spectrosc.*, **30**, 301 (1999).
25. J. Zuo, C. Xu, C. Wang and Z. Yushi, *J. Raman Spectrosc.*, **30**, 1053 (1999).
26. M.L. Dele, P. Dhamelincourt, J.P. Poroit and H.J. Schnubel, *J. Mol. Struct.*, **143**, 135 (1986).
27. R.J.H. Clark, M.L. Curri and C. Largana, *Spectrochim. Acta*, **53A**, 597 (1997).
28. L.I. McCann, K. Trentleman, T. Possley and B. Golding, *J. Raman Spectrosc.*, **30**, 121 (1999).
29. H.G.M. Edwards, D.W. Farewell and A. Quye, *J. Raman Spectrosc.*, **28**, 243 (1997).
30. H.G.M. Edwards, D.E. Hunt and M.G. Sibley, *Spectrochim. Acta*, **54**, 745 (1998).
31. E.A. Carter and H.G.M. Edwards, in: *Infrared and Raman Spectroscopy of Biological Materials*, H.-U. Gramlich and B. Yan (eds), Marcel Dekker, New York, 2001.

32. J. Chalmers and P. Griffiths (eds), *Handbook of Vibrational Spectroscopy*, vol. 4, John Wiley & Sons, Inc., New York, 2001.

33. P.J. Hendra and J.K. Agbenyega (eds), *The Raman Spectra of Polymers*, Wiley, 1993.

34. B. Schrader, *Raman/Infrared Atlas of Organic Compounds*, 2nd Edition, Wiley-VCH, Weinheim, 1989.

35. A. Garton, D.N. Batchelder and C. Cheng, *Appl. Spectrosc.*, **47**(7), 922 (1993).

36. J.M. Chalmers and N.J. Everall, in: *Polymer Characterisation*, B.J. Hunt and M.I. James (eds), Blackie Academic, Glasgow, 1993.

37. S.W. Cornell and J.L. Koenig, *Macromolecules*, **2**, 540 (1969).

38. J.A. Frankland, H.G.M. Edwards, A.F. Johnson, I.R. Lewis and S. Poshachinda, *Spectrochim. Acta*, **47A**, 1511 (1991).

39. K.D.O. Jackson, M.J.R. Loadman, C.H. Jones and G. Ellis, *Spectrochim. Acta*, **46A**, 217 (1990).

40. K. Tashiro, Y. Ueno, A. Yoshioka, F. Kaneko and M. Kobayashi, *Macromol. Symp.*, **114**, 33 (1999).

41. K. Tashiro, S. Sasaki, Y. Ueno, A. Yoshioka and M. Kobayashi, *Korea Polym. J.*, **8**, 103 (2000).

42. N.J. Everall, J.M. Chalmers, L.H. Kidder, E.N. Lewis, M. Schaeberle and I. Levin, *Polym. Mater. Sci. Eng.*, **82**, 398–399 (2000).

43. H.-J. Sue, J.D. Earls, R.E. Hefner Jr., M.I. Villarreal, E.I. Garcia-Meitin, P.C. Yang, C.M. Cheetham and C.J. Plummer, *Polymer*, **39**, 4707 (1998).

44. J.R. Walton and K.P.J. Williams, *Vib. Spectrosc.*, **1**, 239 (1991).

45. K.E. Chike, M.L. Myrick, R.E. Lyon and S.M. Angel, *Appl. Spectrosc.*, **47**, 1631 (1993).

46. M. Kawagoe, M. Takeshima, M. Nomiya, J. Qiu, M. Morita, W. Mizuno and H. Kitano, *Polymer*, **40**, 1373 (1999).

47. M. Kawagoe, S. Hashimoto, M. Nomiya, J. Qiu, M. Morita, W. Mizuno and H. Kitano, *J. Raman Spectrosc.*, **30**, 913 (1999).

48. D.L. Gerrard and W.F. Maddams, *Macromolecules* **8**, 55 (1975).

49. A. Baruya, D.L. Gerrard and W.F. Maddams, *Macromolecules*, **16**, 578 (1983).

50. E.D. Owen, M. Shah, N.J. Everall and M.V. Twigg, *Macromolecules*, **27**, 3436 (1994).

51. H.E. Schaffer, R.R. Chance, R.J. Sibley, K. Knoll and R.R. Schrock, *J. Phys. Chem.*, **94**, 4161 (1991).

52. J.M. Chalmers and G. Dent, in: *Industrial Analysis with Vibrational Spectroscopy*, Royal Society of Chemistry, London, 1997.

53. I. Persaund and W.E.L. Grossman, *J. Raman Spectrosc.*, **24**, 107 (1993).

54. M. Majoube and M. Henry, *Spectrochim. Acta A*, **47**, 1459 (1991).

55. K. Neipp, Y. Wang, R.R. Desari and M.S. Field, *Appl. Spectrosc.*, **49**, 780 (1995).

56. C. Rodger, W.E. Smith, G. Dent, M. Edmondson, *J. Chem. Soc. Dalton Trans.*, **5**, 791–799 (1996).

57. D. Graham, W.E. Smith, A.M.T. Lineacre, C.H. Munro, N.D. Watson and P.C. White, *Anal. Chem.*, **69**, 4703–4707 (1997).

58. D. Graham, B.J. Mallinder and W.E. Smith, *Angewandte Chemie Int. Ed. Engl.*, **6**, 1061–1063 (2000).

59. D. Graham, B.J. Mallinder and W.E. Smith, *Biopolymers(Biospectroscopy)*, **112**, 1103–1105 (2000).
60. D. Bourgeois and S.P. Church, *Spectrochim. Acta A*, **46**, 295 (1990).
61. N. Everall, *Spectrochim. Acta A*, **49**, 727–730 (1993).
62. G. McGeorge, R.K. Harris, A.M. Chippendale and J.F. Bullock, *J. Chem. Soc. Perkin Trans.*, **2**, 1733 (1996).
63. G. McGeorge, R.K. Harris, A.S. Bastanov, A.V. Churakov, A.M. Chippendale, J.F. Bullock and Z. Gan, *J. Chem. Soc. Perkin Trans.*, **102**, 3505–3513 (1998).
64. P.C. White, C. Rodger, V. Rutherford, W.E. Smith and M. Fitzgerald, *SPIE*, **3578**, 77 (1998).
65. P.C. White, C.H. Munro and W.E. Smith, *Analyst*, **121**, 835 (1996).
66. P.C. White, C. Rodger, V. Rutherford, D. Broughton and W.E. Smith, *Analyst*, **123**, 1823 (1998).
67. J.A.G. Drake (ed.), *Chemical Technology in Printing Systems*, Royal Society of Chemistry, London, 1993.
68. C. Rodger, *The Development of SERRS as a Quantitative and Qualitative Analytical Technique*, Ph.D. Dissertation, University of Strathclyde, Glasgow, 1997.
69. C. Rodger, G. Dent, J. Watkinson and W.E. Smith, *Appl. Spectrosc.*, **54** (2000).
70. D.R. Armstrong, J. Clarkson and W.E. Smith, *J. Phys. Chem.*, **99**, 17825 (1995).
71. K.I. Mullen, D.X. Wang, L.G. Crane and K.T. Carron, *Anal. Chem.*, **64**, 930–936 (1992).
72. H. Zollinger, *Colour Chemistry*, VCH, Weinheim, 1991.
73. K. Venkataraman, *The Analytical Chemistry of Synthetic Dyes*, John Wiley & Sons, New York, 1977.
74. A. Tsumura, H. Koezuka and T. Ando, *Appl. Phys. Lett.*, **49**, 1210 (1986).
75. J.H. Burroughs, C.A. Jones and R.H. Friend, *Nature*, **335**, 137 (1988).
76. Z. Bao, J.A. Rodgers and H.E. Katz, *J. Mater. Chem.*, **9**, 1895 (1999).
77. G. Yu, J. Gao, J.C. Hummelen, F. Wudl and A.J. Heeger, *Science*, **270**, 1789 (1995).
78. J.H. Burroughs, D.D.C. Bradley, A.R. Brown, R.N. Marks, K. Mackay, R.H. Friend, P.L. Burns and A.B. Holmes, *Nature*, **347**, 539 (1990).
79. R.H. Friend, R.W. Gymer, A.B. Holmes, J.H. Burroughs, R.N. Marks, C. Taliani, D.D.C. Bradley, D.A. Dos Santos, J.L. Brédas, M. Lögdlund and W.R. Salaneck, *Nature*, **397**, 121 (1999).
80. T.A. Skotheim, R.L. Elsenbaummer and J.R. Reynolds (eds), *Handbook of Conducting Polymers*, Marcel Dekker, New York, 1997.
81. N.S. Sariciftci (eds), *Primary Photoexcitations in Conjugated Polymers: Molecular Exciton versus Semiconductor Band Model*, World Scientific, Singapore, 1997.
82. H. Keiss (ed.), *Conjugated Conducting Polymers*, Springer-Verlag, Berlin, 1992.
83. Y. Shirota, *J. Mater. Chem.*, **10**, 1 (2000).
84. Aldrich Online Chemical Catalogue, www.sigmaaldrich.com/Brands/Aldrich/Polymer_Products/Specialty_Areas.html.
85. H. Becker, H. Spreitzer, W. Kreuder, E. Kluge, H. Schenk, I. Parker and Y. Cao, *Adv. Mater.*, **12**, 42 (2000).
86. S. Bérnard and P. Yu, *Adv. Mater.*, **12**, 48 (2000).
87. W.P. Su, J.R. Schrieffer and H.J. Heeger, *Phys. Rev. B*, **22**, 2099 (1980).
88. W.P. Su and J.R. Schrieffer, *Proc. Natl. Acad. Sci. USA*, **77**, 5626 (1980).
89. J.L. Brédas, R.R. Chance and R. Sibley, *Mol. Cryst. Liq. Cryst.*, **77**, 253 (1981).

90. M. Fleischmann, P.J. Hendra and A.J. McQuillan, *Chem. Phys. Lett.*, **26**, 163 (1974).

91. M. Yoshikawa and N. Ngai, in: *Handbook of Vibrational Spectroscopy*, J. Chalmers and P. Griffiths (eds), vol. 4, John Wiley & Sons, New York, 2001, p. 2604.

92. M.D. Schaeberle, D.D. Tuschel and P.J. Treado, *Appl. Spectrosc.*, **55**, 257–266 (2001).

93. F. Cerdeira, T.A. Fjeldly and M. Cardona, *Phys. Rev. B*, **8**, 4734 (1973).

94. M. Yoshikawa, K. Agawam, N. Morita, T. Matsunobe and H. Ishida, *Thin Solid Films*, **310**, 167 (1997).

95. J.-H. Kim, S.-H. Seo, S.-M. Yun, H.-Y. Chang, K.-M. Lee and C.-K. Choi, *Appl. Phys. Lett.*, **68**, 1507 (1996).

96. S. Nakashima and H. Harima, in: *Handbook of Vibrational Spectroscopy*, J. Chalmers and P. Griffiths (eds), vol. 4, John Wiley & Sons, Inc., New York, 2001, pp. 2637–2650.

97. F.H. Pollak and R. Tsu, *Proc. SPIE*, **452**, 26 (1984).

98. S. Nakshima and M. Hangyo, *Trans. IEEE*, **QE-25**, 965 (1989).

99. P. Dhamelincourt and S. Nakshima, in: *Raman Microscopy*, G. Turrel and J. Corset (eds), Academic Press, London, 1996.

100. H.-U. Gramlich and B. Yan (eds), *Infrared and Raman Spectroscopy of Biological Materials*, Marcel Dekker, 2001.

101. P.T.C. Freire, in: *Proceedings of the International Conference on Raman Spectroscopy*, S.L. Zhang and B.F. Zhu (eds), John Wiley & Sons, 2000, p. 440.

102. K.C. Schuster, I. Reese, E. Urlab, J.R. Gapes and B. Lendl, *Anal. Chem.*, **72**, 5529 (2000).

103. T.A. Alexander, P.M. Pelligrino and J.B. Gillespie, *Appl. Spectrosc.*, **57**, 1340–1345 (2003).

104. J.P. Wold, B.J. Marquardt, B.K. Dable, D. Robb and B. Hatlen, *Appl. Spectrosc.*, **58**, 395–403 (2004).

105. G.D. Sockalingum, H. Lamfarraj, A. Beljebbar, P. Pina, M. Delavenne, F. Witthuhn, P. Allouch and M. Manfait, *SPIE*, **3608**, 185 (1999).

106. C. Arcangeli and S. Cannistraro, *Biopolymers*, **57**, 179–186 (2000).

107. O. Piot, J.C. Autran and M. Manfait, *J. Cereal Sci.*, **34**, 191–205 (2001).

108. H. Matsi and S. Pan, *J. Phys. Chem. B*, **104**, 8871 (2000).

109. J. Zheng, Q. Zhou, Y. Zhou, T. Lou, T.M. Cotton and G. Chumanov, *J. Electroanal. Chem.*, **530**, 75–81 (2002).

110. M. Manfait, P. Lamaze, H. Lamfarraj, M. Pluot and G.D. Sockalingum, *Biomed. Spec., SPIE*, **3918**, 153 (2000).

111. C.J. Frank, R.L. McCreery and D.C. Redd, *Anal. Chem.*, **67**, 777–783 (1995).

112. E. Atherton, D.L. Clive and R.C. Sheppard, *J. Am. Chem. Soc.*, **97**, 6584 (1975).

113. B. Yan, H.-U. Gremlich, S. Moss, G.M. Coppola, Q. Sun and L. Liu, *J. Comb. Chem.*, **1**, 46–54 (1999).

114. C.-D. Chang and J. Meisenhofer, *Int. J. Protein Res.*, **11**, 246 (1978).

115. E. Atherton, H. Fox, D. Harkiss, J.C. Logan, R.C. Sheppard and B.J. Williams, *J. Chem. Soc. Chem. Commun.*, **537** (1978).

116. J. Ryttersgaard, B. Due Larsen, A. Holm, D.H. Christensen and O. Faurskov Nielsen, *Spectrochim. Acta A*, **53**, 91–98 (1997).

117. D.E. Pivonka, K. Russell and T.W. Gero, *Appl. Spectrosc.*, **50**, 1471 (1996).
118. D.E. Pivonka, D.L. Palmer and T.W. Gero, *Appl. Spectrosc.*, **53**, 1027 (1999).
119. D.E. Pivonka, *J. Comb. Chem.*, **2**, 33–38 (2000).
120. Application Note, In-situ Analysis of Combinatorial Beads by Dispersive Raman Spectroscopy, *Nicolet*, **AN-00121** (2001).
121. G. Ellis, P.J. Hendra, C.M. Hodges, T. Jawhari, C.H. Jones, P. LeBarazer, C. Passingham, I.A.M. Royaud, A. Sanchez-Blazquez and G.M. Warnes, *Analyst*, **114**, 1061–1066 (1989).
122. E.A. Cutmore and P.W. Skett, *Spectrochim. Acta*, **49**, 809–818 (1993).
123. W.C. McCrone, in: *Physics and Chemistry of the Organic Solid State*, D. Fox, M.M. Labes and A. Weissberger (eds), vol. II, Interscience, New York, 1965, p. 275.
124. T.L. Threlfall, *Analyst*, **120**, 2435 (1995).
125. J. Anwar, S.E. Tarling and P. Barnes, *J. Pharm. Sci.*, **78**, 337 (1989).
126. G.A. Neville, H.D. Beckstead and H.F. Shurvell, *J. Pharm. Sci.*, **81**, 1141 (1992).
127. C.M. Deeley, R.A. Spragg and T.L. Threlfall, *Spectrochim. Acta*, **47**, 1217 (1991).
128. A.H. Tudor, M.C. Davies, C.D. Melia, D.G. Lee, R.C. Mitchell, P.J. Hendra and S.F. Church, *Spectrochim. Acta*, **47**, 1389 (1991).
129. S. Paul, C.H.J. Schutte and P.J. Hendra, *Spectrochim. Acta*, **46**, 323 (1990).
130. G. Jalsovszky, O. Egyed, S. Holly and B. Hegedus, *Appl. Spectrosc.*, **49**(8), 1142 (1995).
131. P.J. Hendra, in: *Modern Techniques in Raman Spec*troscopy, J.J. Laserna (ed.), Wiley, 1996, p. 89.
132. Application Note, Polymorph Analysis by Dispersive Raman Spectroscopy, *Nicolet*, **AN119** (2001).
133. C. Frank, in: *Analytical Applications of Raman Spectroscopy*, M.J. Pelletier (ed.), Blackwell Science, Oxford, 1999, pp. 224–275.
134. R. Hilfiker, J. Berghausen, C. Marcolli, M. Szelagiewicz and U. Hofmeier, *Eur. Pharm. Rev.*, **2**, 37–43 (2002).
135. C. Cheng, T.E. Kirkbride, D.N. Bachelder, R.I. Lacey and T.G. Sheldon, *J. Forensic Sci.* **40**, 31 (1995).
136. W.E. Smith, C. Rodger, G. Dent and P.C. White, in: *Handbook of Raman Spectroscopy*, I.R. Lewis and H.G. Edwards (eds), Marcel Dekker, 2001.
137. N.J. Everall, I.M. Clegg and P.W.B. King, in: *Handbook of Vibrational Spectroscopy*, J. Chalmers and P. Griffiths (eds), vol. 4, John Wiley & Sons, 2001, pp. 2770–2801.
138. I.R. Lewis, in: *Handbook of Raman Spectroscopy*, I.R. Lewis and H.G.M. Edwards (eds), Marcel Dekker, New York, 2001, pp. 919–974.
139. L.S. Plano and F. Adar, *Proc. SPIE*, **822**, 52 (1987).
140. H.C. Tsai and D.B. Bogy, *J. Vac. Sci. Technol. A*, **5**, 3287 (1987).
141. F. Adar, R. Geiger and J. Noonan, *Appl. Spectrosc. Rev.*, **32**, 45 (1997).
142. F.H. Pollack, in: *Analytical Raman Spectroscopy*, J.G. Grasselli and B.J. Bulkin (eds), John Wiley & Sons, New York, 1991, pp. 137–221.
143. I. de Wolf, in: *Analytical Applications of Raman Spectroscopy*, M.J. Pelletier (ed.), Blackwell Science, Oxford, 1999, pp. 435–472.
144. S. Nakashima and H. Harima, in: *Handbook of Vibrational Spectroscopy*, J. Chalmers and P. Griffiths (eds), vol. 4, John Wiley & Sons, 2001, pp. 2637–2656.

145. V. Wagner, W. Ritcher, J. Geurtus, D. Drews and D.R.T. Zahn, *J. Raman Spectrosc.*, **27**, 265 (1996).
146. V. Wagner, D. Drews, N. Esser, D.R.T. Zahn, J. Geurtus and W. Ritcher, *J. Appl. Phys.*, **75**, 7330 (1994).
147. D. Drews, A. Schneider, D.R.T. Zahn, D.A. Evans and D. Wolfframm, *Appl. Surf. Sci.*, **104/105**, 485 (1996).
148. D.R.T. Zahn, *Appl. Surf. Sci.*, **123/124**, 276 (1998).
149. J.J. Freeman, D.O. Fisher and G.J. Gervasio, *Appl. Spectrosc.*, **47**, 1115 (1993).
150. G.J. Gervasio and M.J. Pelletier, *At-Process*, **3**, 7 (1997).
151. J.P. Besson, P.W.B. King, T.A. Wilkins, M. McIvor and N. Everall, European Patent Application EP 0 767 222 A2, 'Calcination of Titanium Dioxide' (1996).
152. E. Gulari, K. McKeigue and K.Y.S. Ng, *Macromolecules*, **17**, 1822 (1984).
153. J. Clarkson, S.M. Mason and K.P.J. Williams, *Spectrochim. Acta*, **47A**, 1345 (1991).
154. N. Everall and B. King, *Macromolecules*, **141**, 103 (1999).
155. C. Wang, T.J. Vickers, J.B. Schlenoff and C.K. Mann, *Appl. Spectrosc.*, **46**, 1729 (1992).
156. N. Everall, in: *Analytical Applications of Raman Spectroscopy*, M.J. Pelletier (ed.), Blackwell Science, Oxford, 1999, pp. 127–192.
157. C. Bauer, B. Anram, M. Agnely, D. Charmot, J. Sawatzki, N. Dupy and J.-P. Huvenne, *Appl. Spectrosc.*, **54**, 528 (2000).
158. J.B. Cooper, T.M. Vess, L.A. Campbell and B.J. Jensen, *J. Appl. Polym. Sci.*, **62**, 135 (1996).
159. J.F. Aust, K.S. Booksh, C.M. Stellman, R.S. Parnas and M.L. Myrick, *Appl. Spectrosc.*, **51**, 247 (1997).
160. M.L. Scheepers, J.M. Gelan, R.A. Carleer, P.J. Adriaesens, D.J. Vanderzande, B.J. Kip and P.M. Brandts, *Vib. Spetrosc.*, **6**, 55 (1993).
161. J.B. Cooper, K.L. Wise and B.J. Jensen, *Anal. Chem.*, **69**, 1973 (1997).
162. L. Xu, C. Li and K.Y.S. Ng, *J. Phys. Chem. A*, **104**, 3952 (2000).
163. P.J. Hendra, D.B. Morris, R.D. Sang and H.A. Willis, *Polymer*, **23**, 9 (1982).
164. D.B. Chase, in: *XVth International Conference on Raman Spectroscopy*, S. Asher and P. Stein (eds), John Wiley & Sons, Pittsburgh, 1996, p. 1072.
165. D.B. Chase, *Mikrochim. Acta*, **14**, 1 (1997).
166. S. Farquharson and S.F. Simpson, *Proc. SPIE*, **1681**, 276 (1992).
167. N. Everall, in: *An Introduction to Laser Spectroscopy*, D.L. Andrews and A.A. Demidov (eds), Plenum Press, New York, 1995.
168. N. Everall, B. King and I. Clegg, *Chem. Britain*, **July**, 40 (2000).
169. A.D. Shaw, N. Kaderbhal, A. Jones, A.M. Woodward, R. Goodacre, J.J. Rowland and D.B. Kell, *Appl. Spectrosc.*, **53**, 1419 (1999).
170. D. Renaut, J.C. Pourny and R. Capitini, *Optics Lett.*, **5**, 233 (1980).
171. W. de Groot and R. Rich, *Proc. SPIE*, **3535**, 32 (1999).
172. W.H. Weber, M. Zanini-Fisher and M.J. Pelletier, *Appl. Spectrosc.*, **51**, 123 (1997).
173. A.S. Gilbert, K.W. Hobbs, A.H. Reeves and P.P. Hobson, *Proc. SPIE*, **2248**, 391 (1994).
174. I.E. Wachs, in: *Handbook of Raman Spectroscopy*, I.R. Lewis and H.G.M. Edwards (eds), Marcel Dekker, New York, 2001.

175. D. Uy, A.E. O'Neill, L. Xu, W.H. Weber and R.W. McCabe, *Appl. Catal. B*, **41**, 269–278 (2003).
176. V. La Parola, G. Deganello, C.R. Tewell and A.M. Venezia, *Appl. Catal. A*, **235**, 171–180 (2002).
177. G.J. Hutchings, J.S.J. Hargreaves, R.W. Joyner and S.H. Taylor, *Studies Surf. Sci. Catal.*, **107**, 41–46 (1997).

Chapter 7

More Advanced Raman Scattering Techniques

In previous chapters we have introduced the reader to Raman spectroscopy as most spectroscopists and analysts would use it. We have covered the practical application, the theory, some more advanced applications and many examples. However, modern developments in optics are expanding the opportunities for the effective use of Raman scattering through the construction of devices such as portable spectrometers, mapping and imaging stages and devices for the collection of scattering from nanoscale objects. Further, these developments have led to an improvement in the equipment used in specialist laboratories to measure Raman scattering using more advanced techniques. This makes techniques such as time-resolved methods, Raman optical activity and UV Raman scattering more accessible and has led to an increased interest in multi-photon methods. These include hyper Raman scattering, inverse Raman scattering, and various forms of stimulated Raman scattering such as coherent anti-Stokes Raman scattering. The main reason for the increased interest is that the more complex equipment used for these techniques is confined to a few research laboratories and consequently is not generally available. However, modern optics developments which facilitated the increased use of Raman spectroscopy have also made it more practicable to use the more advanced techniques for specific problems where they have significant advantages. Thus, many readers of this text will not have immediate access to the equipment described later in this chapter, but all spectroscopists should be aware of the possibilities for problems which cannot be solved more simply. It is almost impossible to cover all of the possibilities here, but some key examples of techniques which could grow in importance are described briefly, and these and others are reviewed in [1, 2].

Modern Raman Spectroscopy – A Practical Approach   W.E. Smith and G. Dent
© 2005 John Wiley & Sons, Ltd   ISBNs: 0-471-49668-5 (HB); 0-471-49794-0 (PB)

7.1   FLEXIBLE OPTICS

The use of optical systems to improve different facets of modern life such as communications, precise measurement, digital photography and video imaging has meant that there is an ever increasing range of high quality optical components available to improve and simplify the construction of spectrometers. These include small diode lasers with an ever expanding range of different frequencies and reasonable lifetimes, good fibre optic technology, better filters such as notch filters, holographic gratings and micro-positioning stages for mapping and sensitive detectors.

There are many obvious examples of the influence this has had. Efficient fibre optic coupling means that the separation of a probe head from the spectrometer is now a commonplace design, as described in Chapter 2. The use of small notch filters instead of monochrometers to remove Rayleigh scattering and specular reflectance has reduced the size of commercial instruments, and the use of inexpensive CCD detectors developed for the video camera has enabled less expensive but effective Raman systems to be built.

As an example of the potential, we mentioned in Chapter 2 that we recently used a laser pointer as an excitation source with an admittedly expensive Raman microprobe system to collect and detect the Raman scattering. This was effective because the microscope focussed the low power of the laser pen onto a small spot increasing the effective excitation power at the sample. However, the signal was broadened due to the fact that the laser beam is not as monochromatic as that from a high quality laser. Since Raman scattering is recorded as a shift from one specific excitation frequency, if the excitation line covers a range of frequencies, each peak in the spectrum due to the Raman scattering will also cover a range of frequencies. Thus, broadening of the peaks occurs even before natural line broadening from the molecular processes is taken into account. Thus, a good monochromatic source is essential for the best results. This means that in high quality Raman systems even some diode lasers which can be bought quite cheaply can become expensive as a source since they require the addition of a temperature-stabilized unit. Further depending on the type of laser, there may also be a requirement for an optical feedback loop to prevent a frequency shift caused by light back-scattered or reflected into the laser.

When a microscope or a fibre optic cable is used in the collection system, the size of the image is small and this enables a small detector to be used. Figure 7.1 illustrates a system designed for the collection of Raman scattering which can be held in the palm of the hand. This spectrometer uses a filter to remove Rayleigh scattering and specular reflectance, and a fibre optic cable to couple the probe to the monochromator and CCD detection system shown. Even simpler systems can be constructed if all that is required is total Raman

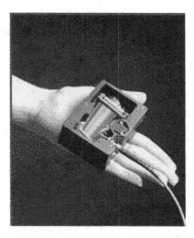

Figure 7.1. Small fibre optic spectrometer. Reproduced by permission of Ocean optics.

scattering rather than a Raman spectrum. In this case, a diode laser, a notch or edge filter and a simple detector such as an avalanche photodiode can be used.

As an example of the advantages of flexible optics, consider the problem of using Raman scattering in liquid chromatography detection. There would be considerable advantages since Raman detection is molecularly specific and would identify the presence of certain analytes without further labelling. However, the lack of sensitivity of Raman scattering in solution is a severe inhibition. The use of a tightly focussed laser beam as is used in a microscope will increase the density at the sample and compensate to some extent. However, the detection limit is often still too high. In Chapter 5 we discussed the possible use of SERS to improve detection limits. However this is often either not applicable or inconvenient. It is also possible to use Raman scattering with other optical arrangements. For example, the laser beam instead of being launched into a fibre optic can be waveguided through a narrow tube filled with a solution containing the substance to be detected. The materials and dimensions for the construction of these tubes require to be correct for effective waveguiding. When set up correctly, the laser beam can pass through many metres of the sample exciting Raman scattering as it passes. It then exits at the far end of the waveguide along with the associated Raman scattered radiation. This huge path length significantly increases the signal to noise ratio of the Raman scattering.

The increased flexibility with fibre optic coupling has led to the combination of Raman spectroscopy with many other techniques. The simplest example does not need a fibre optic coupled head though it can be convenient to use one. Thin layer chromatography (TLC) plates have been examined by Raman spectroscopy to determine the position of spots across the plate. The spots are easily examined, *in situ*, using a microscope or a fibre optic plate reader. The separated organic analytes are easily distinguished by Raman scattering from the inorganic silicates or other material of the chromatography plate. However, it is often the case that fluorescent stains are used with TLC for detection and this can totally obscure the Raman spectrum. In some cases, this can be turned to advantage using resonance or SERRS.

Many other combined techniques which use either fibre optic coupled Raman probes or modified microscope systems with different chromatographies (HPLC, CE, GPC, FIA and GC) have been reported using both on-line and off-line detection. Raman spectrometers have also been coupled to differential scanning calorimeters (DSCs) to follow changes in the spectrum with temperature and to electrochemical cells to follow changes with voltage. In the latter field, optically transparent thin layer electrodes (OTTLE) are often used in electronic spectroscopy to obtain spectra at controllable potentials. Using the extra enhancement of resonance, it is also possible to follow changes in such cells using Raman spectroscopy. An example of the Raman spectrum of the radicals produced in an OTTLE cell when a charge transfer material, used in the production of light emitting diodes, is dissolved in solution is shown in Figure 7.2 (see also Section 6.6).

The spectra were obtained on a microscope system with a macro-sampler (see Section 2.7) using a cell which had been designed for electronic spectroscopy. However, a similar result was obtained from a fibre optic coupled probe. The use of the same cell enabled the additional molecular specificity of Raman scattering to be used to assist the interpretation of the results obtained with electronic spectroscopy.

Other examples of the use of flexible optics have been described in earlier chapters and the important techniques of Raman/SNOM, Raman/AFM and Raman /SEM are described later in this chapter. New techniques are appearing regularly. As an example, the use of optical tweezers is an expanding field, largely because the quality and flexibility of modern optics make it simpler. In tweezing, a tightly focussed intense beam of light irradiates a particle. In some particles such as silica particles, the beam passes through the particle and then re-emits. If light enters from the top, the optical path causes repulsive forces to be created at the surface of the particle where the light enters and where it is re-radiated at the foot. The direction of these forces holds the particle in the beam trapping or 'tweezing' it. The trapped particle can then be manipulated by external optics so that it is placed in a suitable position. Raman spectra can be obtained from these particles simply by collecting the scattered light. This system can be set up to give SERS/SERRS by using silver coated particles. However, silver coated particles do not trap well since instead of the light being transmitted through the particle, it is collected by the surface plasmon, moved round the surface of the particle and re-radiated. One method of carrying out SERRS effectively is to lightly coat the particles with silver so that the beam is

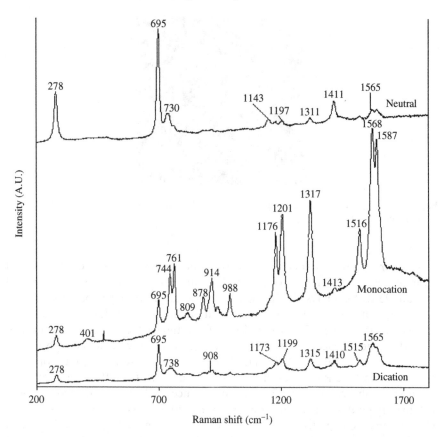

Figure 7.2. Raman spectra taken from an OTTLE cell containing a solution of a charge transfer agent or the mono- or di-cation. The mono and di-cation spectra are resonant or preresonant increasing their intensity. (Reproduced with permission from R. Littleford, M.A.J. Paterson, P.J. Low, D.R. Tackley, L. Jayes, G. Dent, J.C. Cherryman, B. Braon and W.E. Smith, *Phys. Chem., Chem. Phys.*, **6**, 3257–3263 (2004).)

still transmitted but with enough silver on the surface to give SERRS. Figure 7.3 shows SERRS from a lightly coated particle trapped in a tweezer system using 532 nm excitation. The scattering is from points on the particle presumably where the silver is most heavily deposited. One remarkable feature of this is the fact that it can be viewed through a microscope with a standard video camera. This indicates the high sensitivity of SERRS and the consequent ability to detect it with simple equipment. The high power of the tweezing beam can cause ready sample photodegradation and another way to use this system is to use more heavily coated particles which do not trap well and touch the edge of the particle with the beam. This still gives good SERRS but the signal is stable for much longer.

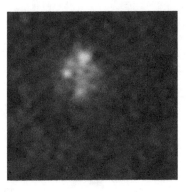

Figure 7.3. SERRS from a single silica particle indicating the short lived pulse of Raman scattering.

Another effective method is to use the evanescent field created when a light beam is close to a surface. In one simple arrangement, a prism is set up so that the sample to be analysed is presented as an adsorbate layer on top of a flat surface and the exciting beam is directed at an angle from below. If the angle of the laser beam is such that it is reflected from the surface, the evanescent field created at the surface causes an electric field on the area directly above the surface where the adsorbate is placed. This will cause Raman excitation from the adsorbate. The Raman excitation is then collected from the side away from the prism as shown in Figure 7.4. However, since the beam is reflected and does not directly contact the adsorbate very much, higher laser powers can be used with less interference from non-Raman scattered light. A more sophisticated development of this is to use a quartz crystal in a manner analogous to that used for ATR in infrared scattering (see Section 2.6.3).

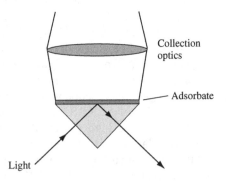

Figure 7.4. Raman scattering collected using evanescent field excitation.

Thus, the modern spectroscopist has a 'toolkit' of components which enable the construction of instruments designed to meet specific problems. The power of this approach is rapidly increasing. One example of this is the handheld 'white powder detector' used in drug detection and described earlier. Section 7.2 describes more advanced ways in which optics and equipment can be configured to improve the spacial resolution of the system.

7.2 TUNEABLE LASERS, FREQUENCY DOUBLING AND PULSED LASERS

With the exception of one example, all the work described so far can be achieved with continuous wave (CW) lasers which continuously emit light of a specific frequency. However, both tuneable and pulsed lasers can have specific advantages in Raman scattering and are widely available. Much of the physics for the development of these systems is not new, but what is new is that the systems are more reliable and simpler to use than was the case a few years ago.

Some laser systems provide tuneable radiation so that the frequency of the monochromatic beam can be selected by the operator within a specific wavelength range. The standard systems most widely used to achieve this are dye lasers or solid state tuneable lasers. In the dye laser, a powerful monochromatic laser beam from a 'pump' laser is passed through a flowing stream of dye solution. The dye is excited and emission occurs over a range of frequencies. A tuned optical cavity is used to create the laser radiation. One frequency is selected, for example by turning a prism in the cavity so that radiation of only the desired frequency is repeatedly passed back and forward inside the cavity. During this process the pump laser continuously excites molecules into the excited state. A process of stimulated emission now occurs. This process is common to all laser systems. When a photon of light passes a molecule containing an electron in an excited state of exactly the same frequency, emission is stimulated and the two photons emerge with the same energy and in phase. Stimulated emission is a relatively efficient process compared to spontaneous emission and consequently molecules pumped up to the excited state by the pump beam preferentially reradiate with the chosen frequency. This creates an amplified beam in the optical cavity in which all the photons are of one frequency and in phase. The beam continues to oscillate back and forward inside the cavity until a power level is reached at which one of the mirrors is not sufficiently strong to prevent the beam passing through it. The laser emission then exits though this mirror. By simply altering the angle of the prism in the cavity, the operator can change the frequency of laser emission within the region in which the particular dye used is effective. Overall this is an inefficient process with considerable energy being lost, for example from all the light which was emitted by the dye and not tuned into the cavity, but powerful pump lasers are readily available and consequently an effective system can be purchased or constructed. The main problem with this system is the use of the flowing stream of dye. More recently, solid state systems have been developed which will provide tuneable radiation. They use ion-doped crystals such as titanium in sapphire in place of the dye. Again, the actual process is inefficient but it is a price worth paying if tuneable radiation is required.

Another development which is now readily available is the use of frequency doubled lasers where a powerful laser such as a Nd/YAG system which emits at 1064 nm is used. The laser beam is directed into a crystal which produces second harmonic generation. The beam of light is emitted at double the frequency but with a much lower power. In addition to frequency doubling, frequency tripling and frequency quadrupling are achievable through the same effect. Thus, a Nd/YAG laser which emits at 1064 nm can be used to obtain appreciable laser power at 244 nm in the UV. Further, a powerful argon ion laser can be frequency-quadrupled to get a range of frequencies in the UV. This provides effective long-lifetime UV sources for Raman scattering.

Finally, there is a very large group of lasers which are used in many optics laboratories and which have been neglected so far. Pulsed laser systems are now readily available in anything from large laser form to chip form. For the study of some processes such as very fast events, these lasers are essential. We use a pulse system when discussing the advantage of Kerr-gating in reducing fluorescence later in this chapter. Other examples of the use of pulsed lasers are given below. The frequency at which pulses are emitted varies from the microsecond to the femtosecond and each frequency range has its own unique merits. The peak power in the pulse in these lasers can be very high but is delivered only for a short period of time. Important parameters in the selection of a pulsed laser include the frequency of the radiation, the peak power and the repetition rate (rep rate). High peak power and fast rep rates might seem ideal but what this will do very often is cause photodegradation; a compromise using a lower rep rate and a lower peak power is often more effective in Raman scattering. In some cases, low peak powers and fast rep rates are used simply to mimic a CW laser, particularly when an available pulse source is in a frequency region where a simple CW laser is not available. In general, pulsed systems are widely used where they are deemed to have a specific advantage such as in the study of fast reaction processes. With the highest frequency systems, the effect of Heisenberg's uncertainty principle becomes marked. In spectroscopy terms this can be stated as,

$$\Delta E \cdot \Delta t = h/2\pi$$

When Δt is very short as in a femtosecond system, ΔE becomes large making the Raman bands very broad.

7.3 SPATIALLY RESOLVED SYSTEMS

A major limitation for most optical techniques is that the smallest sample which can be located and defined requires to be of about the size of the wavelength of the light. It is possible to detect the presence of smaller sized particles but the definition of the image rapidly decreases and even with two-photon techniques, samples less than about 300 nm in diameter cannot be properly defined using visible excitation. However, visible spectroscopies and in particular Raman spectroscopy give considerable molecular information and use lower frequency, less destructive beams than higher frequency techniques such as electron microscopy. As a result, techniques using Raman scattering which can give information from nanoscale samples are of considerable use.

There are simple ways that a good Raman spectrum can be obtained from nanoscale samples. For example, if a particle to be analysed can be isolated and adsorbed on the surface so that no other effective Raman scatterer lies within a few microns of it, Raman scattering can be recorded under a microscope. The sample can then be relocated under a transmission electron microscope (TEM) or scanning electron microscope (SEM) and defined. It has to be recognized that the Raman scattering is measured from only part of the area illuminated by the focussed laser beam. This means that a strong Raman scatterer is more effective here and there can be a problem with increased reflection and scattering from the area of the surface which is irradiated with the focussed radiation but does not contain the sample.

In Chapter 2, Figure 2.19 shows a SERRS map of 30 nm silver particles. The individual Raman signals can be clearly seen with a resolution of 1 μm. However, since all the signals arise from the 30 nm particles, it can be said that the resolution is essentially 30 nm. To locate these particles accurately would require another technique such as TEM/SEM or the use of the atomic force microscope (AFM).

The need to examine structure at a greater resolution than can be obtained optically led to the creation of a combined SEM/Raman system. The Raman beam is introduced into the SEM through a fibre optic probe. It is focussed through a hole in a concave mirror directly onto a sample mounted on the sample stage. The mirror also has a hole to allow the electron beam to contact the sample. The remaining mirror surface is then used to collect a cone of Raman scattering back through the probe for analysis outside the SEM. Figure 7.5 gives a diagram of the equipment. The use of such equipment has considerable potential but care has to be taken with the power of the electron beam. The power often used in a normal thermionic SEM source can destroy organic layers very quickly, so low power settings or a field emission source should be used wherever possible.

Another method that is growing in popularity is the use of scanning near-field optical microscopy (SNOM). In one form of this technique, a glass fibre is

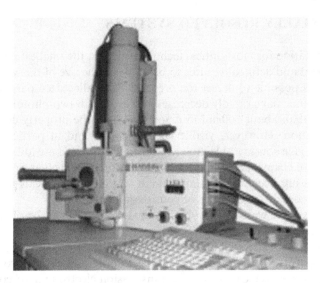

Figure 7.5. A combined SEM/Raman system. The Raman spectrometer is integrated into the side of the SEM. (Reproduced with permission from Renishaw plc.)

coated with aluminium or another metal. The fibre is heated and pulled to provide a very narrow metal clad fibre over a short length. It is then cleaved at the narrow part to leave a narrow, optically clear small aperture. When light is launched down the other end of the fibre, it is contained in the fibre by the metal coating. The amplitude of the light is compressed within the tube as it narrows and the light emerges from the narrow aperture. Apertures down to about 50 nm are used. Consequently, if the fibre can be placed almost in contact with the surface, the effective irradiation area of the surface is a 50 nm circle and this is essentially the spatial resolution of the technique. The way in which this is done is usually to adapt the technology developed for AFM. The very small excitation volume means that fewer molecules are excited compared to normal microscopic Raman techniques. In one sense this can be regarded as an improvement in sensitivity. However, the inefficiency of the excitation method and the difficulty of collection from close to the tip mean that long accumulation times are often required. The simplest method to collect the Raman scattering is to arrange collection optics to collect a cone of scattering from the small area excited by the beam emerging from the fibre. There is no need for the collection optics to be focussed down to 50 nm since the area irradiated with the intense light will give the most efficient Raman scattering. There are many other possible arrangements of this type but all have essentially the same advantages and disadvantages. The obvious advantage is in spatial resolution, and the disadvantage is that the process of confining the light is inefficient and there are compromises with the collection efficiency as well. The result is that this works best with strong Raman scatterers and with long accumulation times. Raman scattering is also generated in the fibre and must be removed with a notch or edge filter to obtain the best results.

In another technique the AFM is combined with SERS/SERRS detection. Here, the cantilever tip of the AFM is coated with silver or gold so that the tip is covered in metal. If the tip contacts the surface and the excitation beam irradiates the area, strong SERS/SERRS can be obtained from the surface. Since this means an enhancement of 10^6 or more, it is light from the point where the tip touches the surface that is collected. The AFM map defines the exact position at which the Raman scattering is collected. Thus, the resolution is approximately that of the tip and certainly very small areas can be located and interrogated in this way. Once again, the Raman collection optics are set to collect a cone of scattered light created by the SERS/SERRS from the tip.

There are many variations on the optical arrangements for these techniques and in fact SNOM is also used with a silver coating so that the edge of the SNOM has a roughened silver tip. In this way, 50 nm Raman excitation is passed directly onto an area of the surface which can be contacted by the silver to create SERS/SERRS. The main drawbacks of these techniques are the compromises made in setting up the instruments, the limited number of samples that are suitable and the difficulties in creating effective, reliable and reproducible coated AFM tips.

7.4 NONLINEAR RAMAN SPECTROSCOPY

In this book so far, Raman scattering has been described in terms of a single photon event in which the Raman scattering efficiency is linearly related to the laser power. However, what happens at higher power densities if no photodecomposition occurs? In this case more than one photon may interact with a molecule at the same time causing a multi-photon event, the magnitude of which is not linearly related to the laser power. This condition is easily achieved using the high peak power of the pulses from pulsed lasers. In addition it can be arranged using pulses from one or more lasers to irradiate the sample at the same time. A number of different forms of Raman scattering have been observed by using this approach, and collectively, these are all nonlinear spectroscopies. They have specific advantages for the solution of specific problems. For example, coherent anti-Stokes Raman scattering (CARS) is one way of reducing fluorescence in highly fluorescent media and hyper Raman spectroscopy has selection rules such that asymmetric bands are relatively more intense in the spectrum.

These techniques can add to our understanding of the nature of the molecule but to date, such systems have been expensive and complicated to build.

Recently, advances in optics have made these processes somewhat more accessible, and although still expensive and complex, they are now becoming available to more spectroscopists. In this book, only the outline of a few of these techniques will be given in order to illustrate the potential of this type of technology.

In hyper Raman spectroscopy, an intense beam of radiation is focussed onto the sample. This is usually achieved using a 1064 nm Nd YAG laser. If sufficient power is present, and two photons interact with the one molecule then a virtual state is created at double the frequency of the laser excitation. Raman scattering from this virtual state to an excited vibrational state of the ground state is called hyper Raman scattering. There could be some advantages in this. Firstly, Raman scattering from a 1064 nm laser is too low in frequency to be detected using a CCD camera and usually requires an interferometer and FT spectrometer. For those laboratories equipped only with systems which use CCD detectors, the hyper Raman effect could be, in principle, a useful way of obtaining Raman scattering with this low frequency laser. Besides providing a different spectrum, this will greatly reduce fluorescence. The main disadvantage is that hyper Raman scattering is very weak, and very high laser powers are required to achieve effective scattering. This is usually achieved with pulsed lasers. Such an arrangement is likely to cause an unacceptable amount of heating and sample degradation in many systems. A diagram of the hyper Raman process is given in Figure 7.6.

Perhaps the most widely used nonlinear technique is CARS. Reviews on this technique can be found in [1, 2]. In this technique, a number of laser excitation sources are required. In its most standard form, three different laser sources are used. One beam creates a virtual state as for ordinary Raman scattering. The

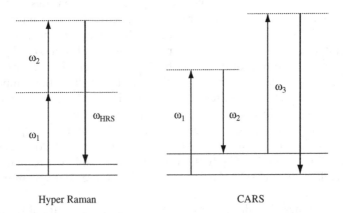

Hyper Raman   CARS

Figure 7.6.   Diagramatic representation of hypho Raman scattering and CARS.

frequency of the second beam is chosen to have a frequency equal to that which would be scattered in spontaneous Stokes Raman scattering. This stimulates the creation of an excited vibrational state. A third laser is then used to excite the molecule to a second virtual state. Scattering from this second virtual state, which returns the molecule to the ground state, is called CARS. A diagram of this process is shown in Figure 7.6. This effect can be somewhat simplified by using only two lasers. In this case the photons shown as upward (ω_1 and ω_3) are from the same laser so that $\omega_1 = \omega_3$.

Clearly, it is essential for these nonlinear processes that all photons involved are present on the molecule at the same time. This means that the phase matching of the beams involved is critical. This can be done simply by making the beams co-linear but, until recently, in the most common CARS setup, technique in which the beams were arranged at an angle to each other was used. The reason for doing this is that, if the beams are co-linear, the length over which the scattering occurs can be quite long and consequently it can be quite difficult to collect the radiation effectively from a volume containing enough molecules undergoing the CARS process. When the laser beams are set at an angle, the interacting length between the beams is much shorter making collection easier. However, although this works, the phase matching condition is quite complex and difficult to calculate. Angles between the beams of about 7° are typically used. This type of arrangement is known as BOXCARS. It should be noted that unlike ordinary Raman scattering, CARS is emitted in specific directions and has to be detected in these directions.

Recently, a considerably simplified version of the CARS system has been developed which makes it more practical. In this, the laser light from two lasers is sent co-linearly and in phase down a microscope. The sharp focus of the microscope removes the difficulty with co-linear beams and still provides sufficiently effective phase matching to obtain effective CARS at the focus. This makes these systems much simpler for the average spectroscopist. In addition, the tuneable laser required to provide the second frequency can be easily obtained in a reliable manner using a modern solid state laser and an optical parametric oscillator (OPO). These systems are still expensive and are limited to research applications, but a significant advance in simplicity has been achieved and it may be that in future, further advances will make the technique more widely available.

The main advantage of CARS is that it is an anti-Stokes process and, as a result, fluorescent-free spectra can be obtained. However, in solution, there is an appreciable background associated with CARS that limits the value of this advantage. Again the selection rules are specific to CARS and if the spectrum is compared to the normal Raman or resonance Raman spectrum, this can give a more effective assessment of the properties of a molecule. One example where CARS has been used is in the analysis of gas mixtures in the head of combustion engines.

For many solution phase applications, additional CARS intensity is gained by using excitation frequencies so that resonant or pre-resonant conditions are used. The addition of resonance enhancement gives greater CARS and consequently makes it easier to discriminate the signals from the associated background signal. An example of CARS for rhodopsin is shown in Figure 7.7.

Both hyper Raman and CARS are often used with excitation frequencies which are close to that of an electronic transition in the system. This enables a pre-resonant or resonant effect to increase the intensity of the signal. Clearly, SERS will be very effective with these methods as well. Figure 7.8 shows a spectrum of hyper Raman/SERS. The compound chosen here (pyrazine) is one that has often been used to probe the SERS effect. The main advantage is that it has a centre of symmetry and consequently in Raman scattering could give symmetric vibrations with no evidence of any infrared active vibrations in the spectrum. In SERS there are more vibrations since the adsorption to the

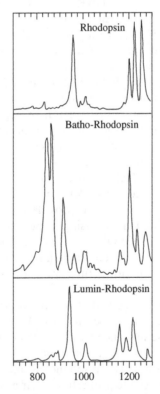

Figure 7.7. CARS of rhodopsin. (Reproduced with permission from F. Yager, L. Ujj and G.H. Atkinson, *J. Am. Chem. Soc.*, **119**, 12610 (1997).)

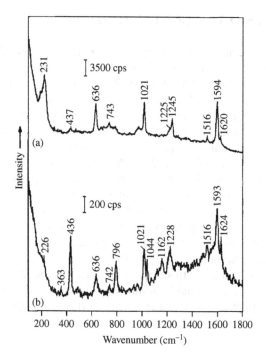

Figure 7.8. SERS (top) and surface enhanced hyper Raman scattering (foot) of pyrazine. (Reproduced with permission from W.H. Li, X.Y. Li and N.T.U. Yu, *Chem. Phys. Lett.*, **305**, 303 (1999).)

surface breaks the centre of symmetry and makes it possible for infrared active bands to appear. In surface enhanced hyper Raman, even more breakthrough of infrared active modes is observed (see Figure 7.8).

There are many more ways in which pulsed lasers can be combined to achieve nonlinear Raman scattering. Stimulated Raman scattering and inverse Raman scattering are two of the most commonly cited. Stimulated scattering is created by tuning in a second frequency to stimulate the Raman scattering process in a manner analogous to that used to create CARS but using only ω_1 and ω_2. In inverse Raman scattering both the Rayleigh and Raman scattering wavelengths are excited simultaneously. In some cases broadband radiation is used to excite the Raman bands. In the correct circumstances energy transfer within the molecule can lead to absorption at the Raman frequencies rather than scattering, hence the use of the term 'inverse scattering'. However, these techniques are little used and a considerable amount of spectroscopic theory underlies the exact nature of the effects. The *Handbook of Vibrational Spectroscopy* [1] has short articles which would act as a lead-in to further studies in this field.

7.5   TIME RESOLVED SCATTERING

With CW lasers, conventional Raman scattering is obtained by collecting photons over a period of time which can extend from seconds to hours, depending on the sample. To obtain evidence on events which occur in the pico- or nanosecond timescales using Raman scattering, a number of strategies are used. One of the most common is to use a pump/probe system. In this arrangement, the output from a pulsed laser is divided into two beams. One is delayed by a chosen number of pico-, nano- or femtoseconds. The pump beam then initiates a process such as a photochemical process in the molecule. The delayed probe beam then interrogates the sample and the Raman scattering resulting is recorded. Obviously, the number of photons scattering for any one pulse is usually quite small. By using a detection system which is synchronized with the laser pulses, an appreciable signal is accumulated over many pulses. One problem with this is that the high peak powers can initiate photodegradation as well as the photochemical reaction desired.

One beautiful example of this approach has been the study of the photodissociation of carbon monoxide bound to the heme in certain enzymes. An example of this effect is shown in Figure 7.9. In this approach, the pump beam is used to cause the initial photodissociation. By selecting different delay times

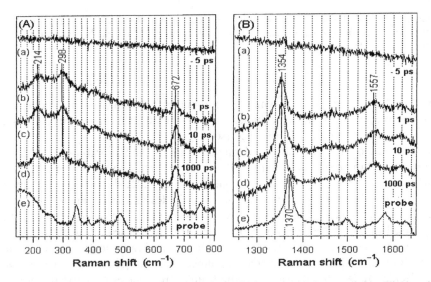

Figure 7.9.   The photodissociation of CO from the heme centre in myoglobin. The band at 1370 cm^{-1} is the oxidation state marker. Dissociation of the CO from the heme causes reduction and other changes. (Reproduced with permission from A. Sato, Y. Sasakura, S. Sugiyama, I. Sagami, T. Shimizy, Y. Mizutani and T. Kitagawa, *J. Biol. Chem.*, **277** (36), 32650–32658, Sep. 6, 2002.)

between the pump and the probe, Raman scattering from carbon monoxide, which occurs at different times after band dissociation and therefore at different separation distances between the carbon monoxide and the heme, can be recorded. From this data, good evidence about the pathway of carbon monoxide desorption in the enzyme is obtained. This is only one of many examples of the use of pulse laser sequences.

Usually, pico- or nanosecond pulses are used to allow the photochemical event to occur and the Raman scattering to be collected before the next pulse. With femtosecond lasers, the pulses occur at a faster rate than the vibrational process (in the picosecond range). Thus, a molecule can be excited many times before one vibration takes place and the excitation is faster even than the Raman scattering process. The result is that a ringing effect can be obtained like hitting a bell with a hammer. The interpretation of the complex data obtained from this type of approach lies outside the scope of this book. However, the information which can be obtained for specific problems particularly concerning short lived species is unique. Systems such as haemoglobin and the light-harvesting proteins have been studied.

Another use of pulsed lasers is to overcome fluorescence, a major disadvantage of Raman scattering, by employing a Kerr gate. The method exploits the fact that when a powerful beam of light is passed through a sample, it can cause changes in the dielectric properties of the medium. This effect, the optical Kerr effect, can be set up so that in an effective medium, the plane of polarization of the incident light is rotated by 90° when radiation passes through it.

By splitting the pulse from the laser into two parts, the system can be set up so that the scattered light from one part of the pulse excites the sample at the same time as the Kerr medium is perturbed by the other part of the pulse. Any light passing through it from the sample has the plane of polarization rotated by 90° in the medium. To create a Kerr gate, the scattered radiation from the sample is passed through a polarizer, into the Kerr medium, then through a second polarizer and onto the detector. The two polarizers are set to be at 90° to each other so that no light can pass from the sample to the detector. This is the closed state of the Kerr gate and is the case before the radiation contacts the sample. However, if a sample is excited by one part of the pulse and the second part passes through the Kerr medium, it will rotate the plane of polarization of the scattered light by 90° so that radiation can pass through the analyser and be detected. This is the open gate. The length of time the gate is open is defined by the shaping of the second part of the pulse so that it arrives slightly later than the scattered light and ends a pre-decided amount of time later.

If Raman scattering and fluorescence are both produced from the sample, no signal will be detected before the Kerr gate is opened by the pulse reaching the Kerr medium or after it is closed by the pulse leaving it. Since Raman scattering is faster than fluorescence in many but not all cases, it is possible to set this system up so that it collects only the first few picoseconds of emission following

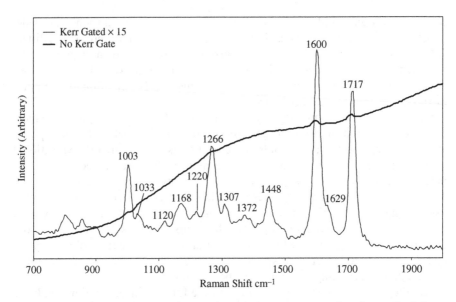

Figure 7.10. Kerr gated (light line) and normal Raman scattering (heavy line) for a fluorescent street sample of cocaine. (Reproduced with permission from R.E. Littleford, P. Matousek, M. Towrie, A.W. Parker, G. Dent, R.J. Lacey and W.E. Smith, *Analyst*, **129**, 505–506 (2004).)

the pulse. This is predominantly Raman scattering. In this way, efficient fluorescence rejection is obtained every time a pulse of light irradiates the sample. These pulses are accumulated over a period of time. Figure 7.10 shows the emission spectrum from a street sample of cocaine, which normally fluoresces, with and without the Kerr gate. The sharp Raman spectrum obtained with the Kerr gate is quite clear. In this case, it is the spectrum of a 75% sample of cocaine. However, some of the weaker bands are actually due to impurities present in the sample which also give effective Raman scattering. These can be seperated by considering the energies and intensities of the normal pure cocaine Raman spectrum.

7.6 RAMAN OPTICAL ACTIVITY

Earlier in this book we discussed mainly the use of CW lasers to produce the excitation beam for Raman scattering. The light used was linearly or plane polarized. However, circularly polarized light can also be used to obtain Raman scattering for molecules that contain a chiral centre. A beam of plane polarized light generated by a laser and a polarizer is passed through a device such as a photoelastic modulator to create circularly polarized light. This

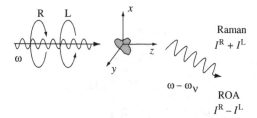

Figure 7.11. Schematic diagram of the basic ROA experiment which measures a small difference in the intensity of Raman scattering from chiral molecules in right (I^R) and left (I^L) circularly polarized light. Incident photons have angular frequency ω and Raman scattered photons have angular frequency $\omega - \omega_v$. The conventional Raman intensity is given by the sum $I^R + I^L$ and the ROA intensity by the difference $I^R - I^L$. For achiral molecules the ROA intensity is zero. Reproduced with permission from E Blanch (UMIST). (Diagram supplied and drawn by Ewan Blanch at UMIST.)

consists of a block of quartz cut at a particular angle so that when linearly polarized light is passed through it and the block is put under stress, the plane of polarization of the light is rotated. The quartz is held by two clamps which oscillate in such a way that a beam is created which alternates between left circularly polarized (lcp) and right circularly polarized (rcp) light. When this light interacts with a chiral molecule, more scattering will occur from either the lcp or rcp beam depending on which fits with the helix which can be traced on the chiral molecule. The difference in intensity between these beams is Raman optical activity (ROA). The selection rules are more complex because the equations used to predict it involve both the electric and magnetic dipole operators. Until recently the very weak signals obtained and the requirement to retain the circular polarization of the beam have made it difficult to exploit the method. However, advances in optics have made it possible for a commercial instrument to be constructed and good spectra can now be obtained in short periods of time. This technique is proving very effective in that not only does it provide information on chirality within amino acids and polypeptides, it will also selectively identify particular features within a protein such as the degree of folding occurring within it. Further, in addition to advances considered earlier for Raman spectroscopy in general, the use of fast effective modulators to achieve this circular polarization has been a key development. A diagram of the technique is shown in Figure 7.11.

7.7 UV EXCITATION

The fourth power nature of the scattering makes UV Raman scattering much more sensitive than visible Raman scattering. Further, the UV region contains

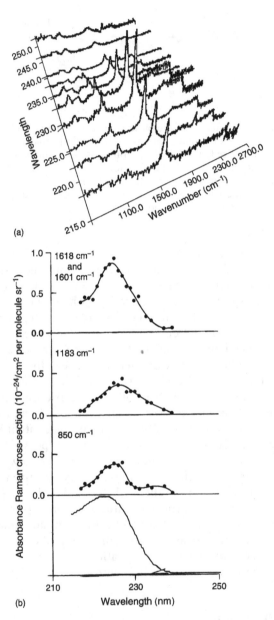

Figure 7.12. UV-resonance Raman scattering for tryptophans. Top: Actual spectra with different excitation wavelengths. Foot: Intensity dependence on excitation wavelength of individual peaks. (Reproduced with permission from M. Ludwig, S.A. Asher, *J. Am. Chem. Soc.*, **110**, 1005–1011 (1988).)

many naturally occurring chromophores in a range of molecules. Thus, UV resonance Raman scattering can be obtained from a larger range of species without the use of a label. Fluorescence is no longer a problem at this very short wavelength. In most systems, there are enough vibronic states for the energy to be dissipated into the material and even if it is emitted for some reason, the emission is well outside the region used for Raman detection. There is a drawback, however, in that the high energy radiation and the presence of many chromophores make photodecomposition an even more serious issue than with visible excitation. Samples are often spun or presented to the instrument in a flowing cell to reduce this problem (see Chapter 2). Using part of the advantage gained from the increased efficiency to use a lower excitation power can also reduce the problem. UV Raman scattering has some unique advantages particularly for biological systems. With the correct conditions and laser frequency, it is possible to obtain molecular resonance from groups such as tryptophans and tyrozines present within a protein. The additional resonance enhancement means that vibrations from the resonant groups can be selectively picked out. Figure 7.12 shows the effect of changing excitation with wavelength for tryptophans. For proteins, the resonance effect can be used to select a specific group. For example, by using a resonant frequency for tryptophans, they can be picked up selectively in proteins such as myoglobin. Changing the pH alters the conformation of the protein. The tryptophans are present in a helix within the protein structure and as the helix opens, the intensity of the bands change.

As the frequency of excitation in the UV is increased, a bigger challenge is created from an equipment point of view. In particular the optics have to be of very high quality. Recently, excitation at 209 nm has been used to obtain resonance Raman spectra from the peptide bond. This work has led to Raman scattering being used to give very specific information on protein structure in an elegant way, which could expand as the quality of inexpensive and simple UV lasers and optics improves.

7.8 CONCLUSIONS

The above techniques are only illustrative of a very much wider range now available in the literature. Optics developments have resulted in a more widespread use of Raman scattering. Examples include portable Raman spectrometers which can be used outside the laboratory, Raman detectors coupled to other instruments and the use of Raman scattering in hostile environments. The Raman spectroscopist has to decide on the nature of the problem faced before deciding whether the more sophisticated techniques outlined later in the chapter are of value. For many problems, Raman scattering using conventional Raman spectroscopy with either visible or infrared excitation will be the simplest choice. If a more advanced technique is required, there is currently a difficulty in that some of the equipment is available only in leading research laboratories and often requires an expert to operate it. However, the worlds of nanotechnology and biotechnology in particular may well create specific needs for the use of the more advanced forms of Raman scattering, and this in turn should lead to wider availability. One technique for which this has recently happened in ROA and CARS microscopes are also becoming more common.

REFERENCES

1. J. Chalmers and P. Griffiths (eds), *Handbook of Vibrational Spectroscopy*, Vols 1 and 2, John Wiley & Sons, Inc., New York, 2001.
2. J.R. Ferraro and K. Nakamoto, *Introductory Raman Spectroscopy*, Academic Press, San Diego, 1994.

Index

Modern Raman Spectroscopy – A Practical Approach W.E. Smith and G. Dent
© 2005 John Wiley & Sons, Ltd ISBNs: 0-471-49668-5 (HB); 0-471-49794-0 (PB)

With kind thanks to Ann Lloyd-Griffiths for compilation of this index.